Lecture Notes in Mathematics 1984

Vicent Caselles · Pascal Monasse

Geometric Description of Images as Topographic Maps

Vicent Caselles
Departament de Tecnologies de la
Informació i les Comunicacions
Universitat Pompeu Fabra
C/Roc Boronat 138
08018 Barcelona
Spain
vicent.caselles@upf.edu

Pascal Monasse
IMAGINE
École des Ponts ParisTech
19 rue Alfred Nobel
77455 Champs-sur-Marne
France
monasse@imagine.enpc.fr

ISBN: 978-3-642-04610-0 e-ISBN: 978-3-642-04611-7
DOI: 10.1007/978-3-642-04611-7
Springer Heidelberg Dordrecht London New York

Lecture Notes in Mathematics ISSN print edition: 0075-8434
ISSN electronic edition: 1617-9692

Library of Congress Control Number: 2009938947

Mathematics Subject Classification (2000): 68U10, 94A08, 05C05

Cover design: SPi Publisher Services

Printed on acid-free paper

springer.com

Preface

This book discusses the basic geometric contents of an image and presents a tree data structure to handle it efficiently. It analyzes also some morphological operators that simplify this geometric contents and their implementation in terms of the data structures introduced. It finally reviews several applications to image comparison and registration, to edge and corner computation, and the selection of features associated to a given scale in images.

Let us first say that, to avoid a long list, we shall not give references in this summary; they are obviously contained in this monograph.

A gray level image is usually modeled as a function defined in a bounded domain $D \subseteq \mathbb{R}^N$ (typically $N = 2$ for usual snapshots, $N = 3$ for medical images or movies) with values in $\mathbb{R}$. The sensors of a camera or a CCD array transform the continuum of light energies to a finite interval of values by means of a nonlinear function g. The contrast change g depends on the properties of the sensors, but also on the illumination conditions and the reflection properties of the objects, and those conditions are generally unknown. Images are thus observed modulo an arbitrary and unknown contrast change.

Mathematical morphology recognizes contrast invariance as a basic requirement and proposes that image analysis operators take into account this invariance principle. An image u is thus a representative of an equivalence class of images v obtained from u via a contrast change, i.e., $v = g(u)$ where g, for simplicity, is a continuous strictly increasing function. Under this assumption, the reliable information in the image is contained in the level sets, be they upper $[u \geq \lambda] := \{x \in D : u(x) \geq \lambda\}$, or lower $[u \leq \lambda] := \{x \in D : u(x) \leq \lambda\}$, independently of their actual level. This theory has been described with detail in the books of Jean Serra, and in the forthcoming book of Frédéric Guichard and Jean-Michel Morel. We review the basic ideas at several points in this monograph, in particular, in Chap. 1.

More recently, taking into account local illumination effects, a more local description of the geometric contents of the image was presented. This led to the introduction of connected components of level sets as basic geometric objects of the images. We shall refer to the description of an image in terms of them as a topographic description of the image. The inclusion between upper (resp. lower) level sets gives an obvious tree structure specific to each family of geometric objects which may be used to encode the image. To be able to handle efficiently both trees in a single data structure which also contains the geometric relations between them, Pascal Monasse was able to fuse both trees into a single one, called the tree of shapes of the image. The basic idea is simple: given a connected component of a level set, be it upper or lower, we fill its holes (an operation that we call saturation). In this way we obtain (in $\mathbb{R}^2$) a simply connected object which can be determined by its boundary and which is called a shape of the image. Since any two shapes of an image are either nested or disjoint, shapes can be structured as a tree, which permits us to handle them in a very efficient way. This is what we call the tree of shapes of an image. We also note that the actual connected components of upper and lower level sets can be recovered from the shapes. To be able to proceed with this construction, we assume that the image is modeled as an upper semicontinuous function. This will be the basic functional model for images adopted in this book. The purpose of Chap. 2 is to introduce the tree of shapes of the image and give its mathematical description. For that we need first to introduce some topological preliminaries and the main properties of the filling saturation operator.

Chapter 3 is devoted to the study of connected operators, which are filters that simplify the topographic map while keeping its essential features. These filters act on the connected components of level sets and simplify the tree of shapes of the images. We shall mainly concentrate on two filters: extrema killers and the grain filter, and we shall study their mathematical properties, and their effect on the tree of shapes of the image. We shall prove that the grain filter is self-dual, i.e., invariant under contrast inversion, and we shall characterize it axiomatically. Let us mention that extrema killers and grain filters can be efficiently implemented using the tree of shapes. We display experiments that illustrate the effect of those filters on images.

The use of a topographic description of images, surfaces, or $3D$ data has been introduced and motivated in different areas of research, including image processing, computer graphics, and geographic information systems (GIS). The motivations for such a description differ depending on the field of application. In all cases these descriptions aim to achieve an efficient description of the basic shapes in the given image and their topological changes as a function of a physical quantity that depends on the type of data (intensity in images, height in data elevation models, etc.).

In computer graphics and geographic information systems, topographic maps represent a high level description of the data. Topographic maps are represented by the contour maps, i.e., the isocontours of the given scalar data. The description of the varying isocontours requires the introduction of data structures, like the *topographic change tree* or *contour tree*, or the Reeb graph, which can represent the nesting of contour lines on a contour map (or a continuous topographic structure). In all cases, the proposed description can be considered as an implementation of Morse theory. Given the scalar data u defined in a domain Ω of $\mathbb{R}^N$ ($u : \Omega \to \mathbb{R}$), the contour map is defined in the literature as the family of isocontours $[u = \lambda]$, $\lambda \in \mathbb{R}$, or in terms of the boundaries of upper (or lower) level sets $[u \geq \lambda]$ ($[u \leq \lambda]$). The first description is more adapted to the case of smooth data (some interpolation will be required in the case of digital data) while the second description can be adapted to more general continuous data where there are plateaus of constant elevation or discontinuous data.

In the context of computer graphics, Morse theory has also been used to encode surfaces in $3D$ space. Several authors have proposed to use a tree structure like the Reeb graph complemented with information about the Morse indices of the singularities and including enough intermediate contours to be able to reconstruct by interpolation the precise way in which the surface is embedded in $3D$ space.

As we already discussed, in image processing, the topographic description was advocated as a local and contrast invariant description of images (i.e., invariant under illumination changes), and its developments have led to the notion of shape and its efficient description in terms of the tree of shapes of the image. The purpose of Chap. 4 is to give the topological description of the tree of shapes and its singularities. For that we introduce and study a topological weak Morse theory for the tree of shapes of the image, and prove its equivalence to other (topological) weak Morse structures which have been used in the literature (let us mention in particular the work of Kronrod). We describe a simple combinatorial algorithm that gives directly the critical levels of the image. In this chapter images are considered to be continuous functions.

Chapter 5 describes the construction of the tree of shapes of an image by fusion of the trees of connected components of upper and lower level sets. Though this algorithm is less efficient when $N = 2$, where some more effective algorithms exist (they are described in Chaps. 6 and 7), it is still useful when $N \geq 3$.

Chapter 6 is devoted to an algorithm computing the tree of shapes of a digital image. The precise data structures and an efficient implementation of the algorithm is given in details. We also study the theoretical complexity of the algorithm.

Chapter 7 is devoted to the description of the tree of bilinear level lines and its algorithmic implementation. After a bilinear interpolation of the discrete data, the image could be treated as a continuous function and a tree of bilinear level lines $[u = \lambda]$ can be computed. Whereas the algorithm uses the same ideas as the one in Chap. 6, there are meaningful differences, the most prominent of which being that there is an infinite number of bilinear level lines, though their structure is finite in nature, as proved in Chap. 4. The tree of bilinear level lines is related to the contour tree computed with the isocontours of the interpolated image. The work of Kronrod can be considered as a mathematical description of the isocontour tree in the case of two-dimensional functions.

Chapter 8 is devoted to applications. Three main categories will be discussed: image comparison, image registration, and extraction of features like edges, corners, or scale adaptive neighborhoods of each pixel. The problem concerning each application is discussed, and we explain how the data structures developed in the present monograph can be applied for an efficient solution of these problems. The experiments illustrate our claims.

Research on this topic has been initiated by the authors roughly 10 years before publication of this monograph. Theoretical and algorithmic advances, so as diverse applications, have been presented at conferences or published in journals by the authors and other researchers, but the present notes regroup them for the first time in a coherent, unified and self-contained framework. We tried to prove all claims by means of elementary results from topology and analysis. We also included numerous simple figures to illustrate the notions and expose various configurations in proofs. We hope that makes them accessible to the general knowledgeable mathematician. Finally, let us mention that reference implementation of most if not all algorithms are included in the open source software suite MegaWave (`http://megawave.cmla.ens-cachan.fr`).

During the preparation of this text V. Caselles was partially supported by the "ICREA Acadèmia" prize for excellence in research funded by the Generalitat de Catalunya and by PNPGC projects, references BFM2003-02125 and MTM2006-14836. Jean-Michel Morel knows that his constant encouragements were a determinant factor in the completion of these notes and the authors are grateful for his support. They would also like to thank Enric Meinhardt for his contributions to Chap. 5. Finally, the feedback from the anonymous reviewers of this monograph was greatly appreciated.

Barcelona, Paris,
July 2009

Vicent Caselles
Pascal Monasse

Contents

List of Figures

List of Algorithms

Chapter 1
Introduction

This chapter is intended as a gentle and casual introduction to the themes developed in these lecture notes. It cites only one reference book for each theme and omits voluntarily all further citations to let it remain an overview of usually independent but here connected topics: Morse theory, topography and mathematical morphology. The three have in common the fundamental role of the (iso)level sets of functions, whose structure is one of the main themes of the present book. The present introduction serves to locate each Chapter into this general context.

1.1 Morse Theory and Topography

In his classical treaty on Morse theory, J. Milnor motivates the subject by discussing the variation of topology of a part of a torus below a plane, as a function of the height of the plane [75]. This reveals the global topology of the torus. Now this analysis can be done also for the graph of a function. The subgraph of a real function defined on the Euclidean plane, called here level set, reveals interest points of the surface, where a slight variation of level changes the structure of the level set.

A customary way to represent the topography on a map is to draw the level lines, that is the lines of constant elevation, as in Fig. 1.1. These level lines, if dense enough, reveal almost all about the terrain they represent. Apart from peaks (local maxima) and pits (local minima), a third category of interesting points emerges as passes (saddle points), which are points where two distinct level lines merge (see Fig. 1.2).

From a mathematical point of view, this analysis can be done with basic differential calculus tools if the function is Morse. A Morse function is twice differentiable and wherever its Jacobian vanishes, its Hessian is invertible. For such well behaved functions, singular points, i.e., points of null Jacobian, are isolated, and the signs of the eigenvalues of the Hessian reveal the nature of the singularity:

V. Caselles and P. Monasse, *Geometric Description of Images as Topographic Maps*, Lecture Notes in Mathematics 1984,
DOI 10.1007/978-3-642-04611-7_1, © Springer-Verlag Berlin Heidelberg 2010

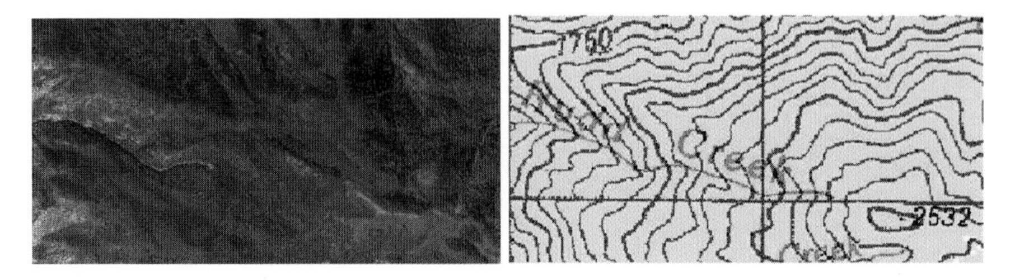

Fig. 1.1 Aerial photograph and topographic map of terrain. Images courtesy the U.S. Geological Survey.

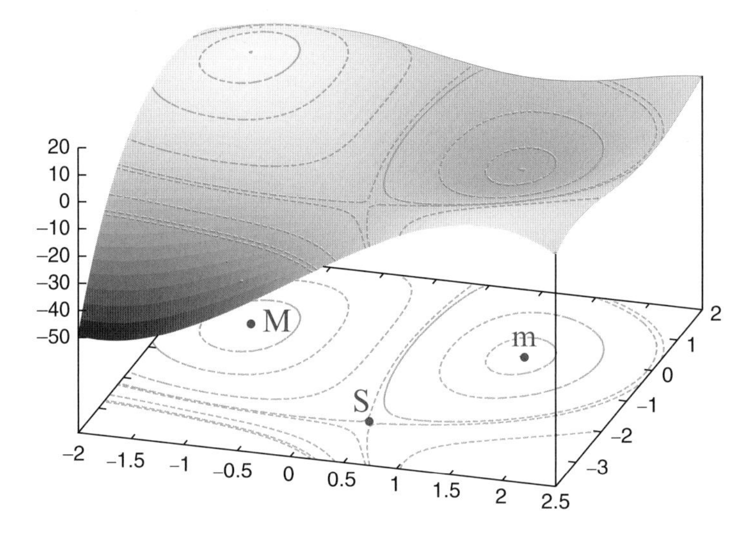

Fig. 1.2 Some level lines and critical points of a Morse function. M is a maximum, m a minimum and S a saddle point.

- 2 positive eigenvalues: a minimum;
- 2 negative eigenvalues: a maximum;
- 2 eigenvalues of different sign: a saddle point.

As our interest lies mostly in image processing, where a gray level image is considered as a function defined on a rectangle of the plane, with the value at a point indicating the amount of light received (see Fig. 1.3), the Morse model is ill suited for our purposes because of one overly optimistic assumption: the invertibility of the Hessian at critical points. Whereas the twice differentiability may be admissible, because a convolution with a Gaussian kernel of small variance would smooth and reduce the effect of noise, the Hessian condition forbids the presence of a plateau in the image. Although an approximation by a Morse function is possible, as Morse functions are dense among continuous functions on a closed rectangle, there is no canonical way

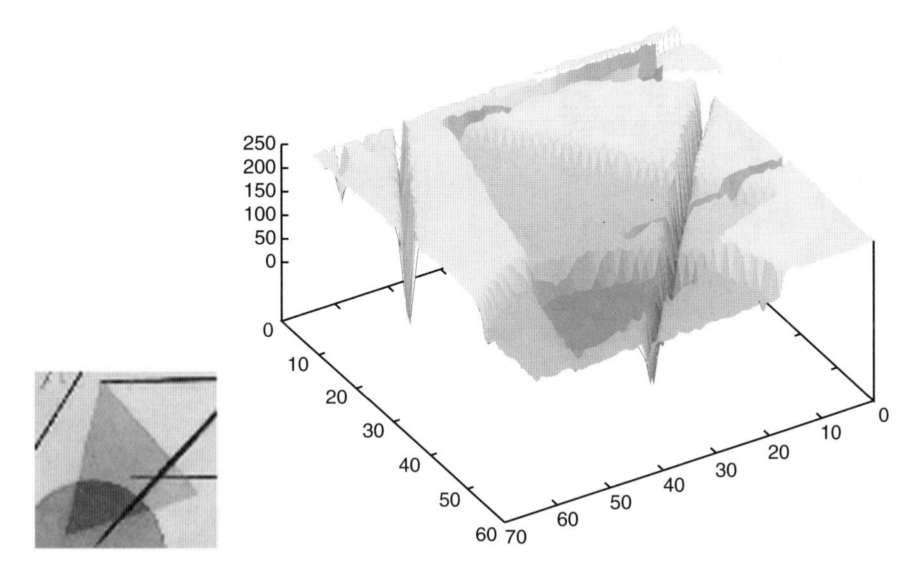

Fig. 1.3 A gray-level image and its representation as a bivariate function graph.

to do this approximation, and the analysis of a Morse approximation does not necessarily reveal much about the topology of the original function.

1.2 Mathematical Morphology

Mathematical morphology, founded by G. Matheron in the 1960s and 1970s, is an image processing theory based on manipulation of sets [69]. The basic operators of mathematical morphology are erosions and dilations. A dilation δ_B by a set B (called structuring element) is an operator that maps a set X to the set $X + B$. Its dual operator is the erosion by the symmetric of B, ε_{-B}, defined by $\varepsilon(X) = \delta(X^c)^c$. Both commute with translations.

If (δ_i) and (ε_i) are families of dilations and erosions, the operators δ and ε defined by $\delta(X) = \bigcap_i \delta_i(X)$ and $\varepsilon(X) = \bigcup_i \varepsilon_i(X)$ are also a dilation and an erosion. In this way, non trivial new operators may be defined. Another way is to combine a dual erosion/dilation. The *opening* $\delta \circ \varepsilon$ and the *closing* $\varepsilon \circ \delta$ are idempotent.

Although originally defined for binary images, mathematical morphology was naturally extended to gray level images by the threshold decomposition and the superposition principle as illustrated by Fig. 1.4. In other words, mathematical morphology operates on level sets of the image, whence the interest of an efficient decomposition of an image into its level sets.

Fig. 1.4 Opening operator by a disk on a gray-level image through threshold decomposition and superposition.

1.3 Inclusion Tree of Level Sets

For a continuously differentiable function, at almost all levels the level lines are Jordan curves. These have a notion of interior and exterior, and as they do not cross, they can be organized in a tree driven by the inclusion order relation: a level line is descendant of another if it is contained in the inner domain of the latter, as in Fig. 1.5. The regular levels do not include in particular levels of extrema, and therefore cannot be directly accounted for in the tree.

In a pioneering work intended to extend the total variation of univariate real functions to functions of two variables, A. Kronrod pointed out that the differentiability has no relevance to such an analysis, and that continuity is sufficient to organize the level sets, defined here as sets of constant level [50]. By defining a kind of quotient topology for the equivalence relation of two points being equivalent if there is a continuum inside a level set joining them, the family of level sets is endowed with a topology making it a dendrite, the topological equivalent of an unrooted tree. Even though the (iso-)level sets are not Jordan curves, they can still be organized in a tree. But there is no root for this tree, in other words there is no *direction* for the dendrite. Actually it can be directed by the choice of a point at infinity.

In Chap. 2, we show that even the continuity condition can be relaxed to semicontinuity, and that the appropriate sets to consider are the upper and lower level sets. These can be organized in a tree because the choice of a point at infinity distinguishes between internal holes and exterior of a level set. Whereas it is straightforward to define a tree of upper, or a tree of lower, level sets, their merging is only enabled by the definition of a point at infinity. The semicontinuity requirement, instead of continuity, is nice to have, because it is compatible with the Mumford-Shah approximation of an image by a piecewise constant image. For example, a piecewise constant model for the image in Fig. 1.3 is obviously better adapted than a global continuous one. Moreover, it fits perfectly the order 0 interpolation of a digital image, that is, the nearest neighbor approximation. This provides the mathematical foundation for the Fast Level Set Transform (FLST), an algorithm that we detail in Chap. 6.

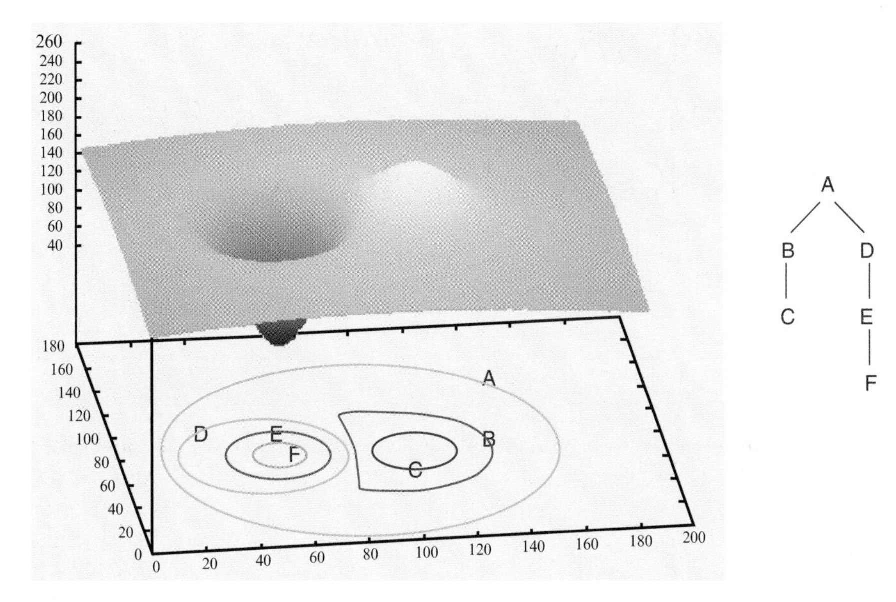

Fig. 1.5 Tree of level lines of a Morse function.

1.4 Topological Description and Computation of Topographic Maps

Going back to the continuity assumption in order to describe in the manner of Morse theory a topographic map, we cannot guarantee that critical points are isolated, and no discrete description of events among level sets is possible. Again, there are many possibilities to approximate the function by another one not suffering from this drawback, but one category stands out as the most natural one: the grain filters. They simply remove small scale oscillations of the function, yielding what will be called a weakly oscillating approximation of the image. Chapter 3 discusses these filters from the point of view of mathematical morphology. One of them treats upper and lower level sets in a symmetric manner and is preferable as it is self-dual in the vocabulary of mathematical morphology (see Fig. 1.6).

Weakly oscillating functions may be analyzed in a discrete description. Chapter 4 discusses several notions that can be defined for such functions: on the one hand the maximal monotone sections (branches of the inclusion tree) and their limit levels (called critical levels), on the other hand the signature of level sets (the family of extrema it contains) and the levels at which the signature changes (called singular levels). These notions are shown to coincide, and when $N = 2$ they generalize to a weakly oscillating function the critical levels of a Morse function. When $N = 3$, the notion of critical value reflects the changes in the number of connected components of isolevel

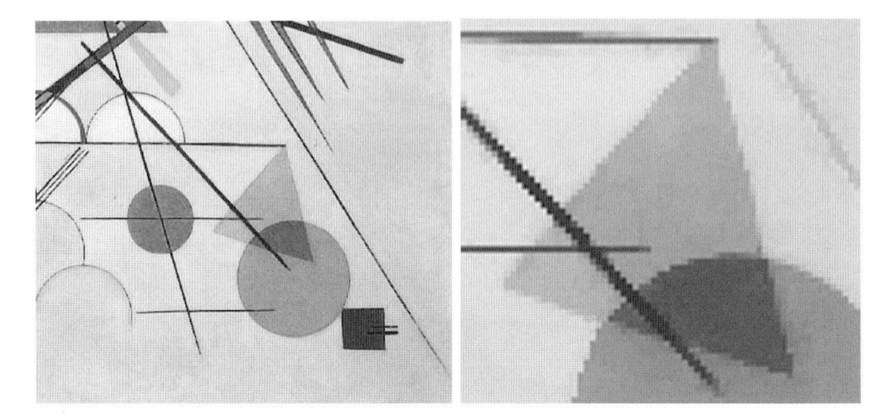

Fig. 1.6 An image before (left) and after (right) application of a grain filter, which removes many small oscillations.

sets $[u = \lambda]$ as $\lambda \in \mathbb{R}$ varies. Let us mention that, in its present form, this theory has not been published elsewhere.

In Chap. 5 we describe an algorithm to construct the tree of shapes by fusion of the trees of connected components of upper and lower level sets. Though the algorithm is less efficient than the algorithm described in Chap. 6, valid when $N = 2$, it is adapted to any dimension.

The considerations of Chap. 4 justify the construction of the tree for an order 1 interpolation of a digital image, i.e., bilinear interpolation. The algorithm to extract it is a variant of the FLST and is presented in Chap. 7. Some applications may use rather the pixelized version for the complete description of the discrete data and its flexibility (invariance to contrast change, among others), while others prefer to use the continuous interpolation for its more regular level lines.

Several applications relying on the inclusion tree have been developed in recent years. We present a few examples of some of them in Chap. 8, ranging from low-level image processing (edge detection, corner extraction) to image alignment and local scale definition.

1.5 Organization of These Notes

The mathematically inclined reader may be most interested in Chaps. 2 to 6, which generalizes the topological Morse description to continuous or semicontinuous functions. Mathematical morphologists may consider more closely Chap. 3 about grain filters, although it is presented in the continuous setting, not in the discrete topology setting most frequent in mathematical morphology.

The computer scientist may focus on Chaps. 6 and 7 for algorithmic considerations, keeping in mind that their full justification are respectively in Chaps. 2 and 4.

All may find motivation for this work in the image processing applications presented in Chap. 8, knowing that their full description must be found among articles referenced in the bibliography.

Chapter 2
The Tree of Shapes of an Image

This chapter presents the tree of shapes of an image, a mix of the component trees of upper and lower level sets. Its existence under fairly weak assumptions and its completeness are proven. Ignoring the small details of the image, we show the essentially finite nature of the tree. Finally, we illustrate these theoretical results with a direct application to gray level quantization.

2.1 Introduction and Motivation

An image, to be processed, must be represented by an adequate model. For example, its Fourier transform is often used for denoising or registration, see [119]. Its decomposition on a wavelet or wavelet packet basis can be used for multiple tasks, the most convincing one being compression, see [63]. These models, totally (in the Fourier case) or partially (in the wavelet case) based on the frequencies of the image, are however dependent on the original image contrast. The same thing can be said of image representations based on edges [66]. While this is not a problem in image compression, it becomes more problematic in image analysis, since we would like any automatic tool to be insensitive to contrast changes, inasmuch as our vision is [118].

In the same way, most algorithms trying to segment the image into significant regions, i.e., trying to find a partition of the image into essentially homogeneous regions, are generally not contrast invariant, see, for instance, [81,82].

Mathematical morphology deals with contrast invariant objects, the most basic ones being the (upper) level sets, $[u \geq \lambda] = \{x, u(x) \geq \lambda\}$, or the level lines $\partial[u \geq \lambda]$ of the image u, [69, 92, 93, 98, 102, 104]. They are said to be contrast invariant, since, for any increasing continuous function g, the level sets of $g \circ u$ are globally the same as the level sets of u, modulo a change of level. Many tasks can be efficiently addressed by means of morphological methods: contrast enhancement [18, 19], filtering [4, 39, 99, 100], compression [35,94], segmentation [72,92,116], intersection [12], or registration [77], to give some examples. Mathematical morphology uses the structure of level sets in

V. Caselles and P. Monasse, *Geometric Description of Images as Topographic Maps*, Lecture Notes in Mathematics 1984,
DOI 10.1007/978-3-642-04611-7_2, © Springer-Verlag Berlin Heidelberg 2010

a fundamental way: indeed, the upper level sets $[u \geq \lambda]$ are nonincreasing with respect to λ, and any contrast invariant process can be interpreted as a geometric process acting on level sets, preserving their order. The resulting image is defined so that its level sets are the processed level sets of the original image (each point gets the level of the smallest processed level set containing it).

As objects of interest can be bright, as well as dark, a frequent requirement of mathematical morphology is to have self-dual algorithms, i.e., algorithms which act in the same manner on upper and lower level sets. And here comes a difficulty, self-dual algorithms modify the families of upper and lower level sets. However, each one of these families yields a complete representation of image, and when we modify one, it is not easy to ensure consistency with the other one.

Moreover, as most objects consist of a single piece, we prefer to take as basic objects of the image the connected components of its level sets and ask the basic operations to act on this structure. Those operations are called connected filters since they preserve connectedness [73, 98, 106]. Connected components of upper (or lower) level sets have a natural tree structure given by the inclusion order. Putting both together creates a self-dual, but redundant, representation of the image. Does a self-dual, and non redundant, representation of the image exist? As this is not possible with both families of connected components of level sets, we shall modify them. The modification used in this paper consists in filling-in its holes. We shall see that, by doing this, we are able to create a single tree of basic objects. This tree structure was first used as an algorithm, but not proved, in [112], and the importance of the role played by holes was not explained.

A very close structure based on level lines, and valid for continuous images, was discovered by Kronrod [50]. However, images have discontinuities, prominently at the edges of objects. The present work can be understood as an extension of Kronrod's tree to semicontinuous images. The tree structure can also be compared to the Morse tree [36, 75], but the regularity required for Morse functions is too stringent for images. An adaptation to images was recently proposed under the name of Digital Morse Theory, [23], but this representation does not show explicitly the upper or lower level sets. The assumption that images are BV functions, i.e., functions of bounded variation [7,24,31], was introduced to deal with the restoration problem [22,30,91,117]. In this model, to define basic objects as connected components of level sets and the corresponding extrema killers, an adapted notion of connectedness, called M-connectedness, is proposed in [6,10]. Apart from the fact that there are clues hinting that images are not of bounded variation due to texture oscillations [3], the M-connectedness will be shown not to yield a tree structure. Thus, we shall restrict ourselves to the case of images as upper-semicontinuous functions, a model which permits discontinuities and the definition of a self-dual tree structure for the shapes of the image. Practical applications of the representation we study here, as well as a fast algorithm to extract it, have been presented in [79, 80]. Let us finally mention that general uses of a tree structure to design filters are explained in [95].

Let us explain the plan of the chapter. Section 2.2 is devoted to define the notions of saturation and of holes, and to study its basic properties. In Sect. 2.3 we prove that the shapes of an image, defined as the saturation of the connected components of its level sets, have a tree structure and we study it. In Sect. 2.4 we prove that the tree is equivalent to the image: the knowledge of the tree is sufficient to reconstruct the image. Section 2.5 shows the finiteness of the tree structure under hypotheses that are met for digital images. Last section is devoted to an application, indeed, we show how to obtain schematic versions of the image by manipulating its associated tree.

2.2 Some Topological Preliminaries

The notation used will be the usual one in topology. If A is a set in a topological space, $\overset{\circ}{A}$, $\overline{A}$ and ∂A will denote, respectively, the interior, the closure and the boundary of A.

Let $\overline{\Omega}$ be a set homeomorphic to the closed unit ball of $\mathbb{R}^N$ $(N \geq 2)$, $\{x \in \mathbb{R}^N, \|x\| \leq 1\}$, and Ω be the interior of $\overline{\Omega}$. Note that, in particular, $\overline{\Omega}$ is compact, connected and locally connected. Moreover, $\overline{\Omega}$ is unicoherent.

Definition 2.1. ([51, §41,X]) A topological space Z is said to be unicoherent if it is connected and for any two closed connected sets A, B in Z such that $Z = A \cup B$, $A \cap B$ is connected.

Let us immediately extend this result to any finite number of sets. The following result repeats Definition 2.1 when $n = 2$ with $A = \overline{\Omega} \setminus X_1$ and $B = \overline{\Omega} \setminus X_2$.

Lemma 2.2. *Let $X_1, X_2, \ldots, X_n$ be disjoint open sets such that for any i, $\overline{\Omega} \setminus X_i$ is connected. Then $\overline{\Omega} \setminus \bigcup_i X_i$ is also connected.*

Proof. Suppose the property proved up to some integer n. Then $X_{n+1} \subseteq \overline{\Omega} \setminus \bigcup_{i=1}^n X_i$, which amounts to

$$(\overline{\Omega} \setminus X_{n+1}) \cup (\overline{\Omega} \setminus \bigcup_{i=1}^{n} X_i) = \overline{\Omega}.$$

The two sets being closed and connected, their intersection, that is $\overline{\Omega} \setminus \bigcup_{i=1}^{n+1} X_i$, is connected. □

The connected components of a set $A \subseteq \mathbb{R}^N$ will be denoted by $\mathcal{CC}(A)$. If $x \in A$, the connected component of A containing x will be denoted by $\mathrm{cc}(A, x)$, and by extension we shall write $\mathrm{cc}(A, x) = \emptyset$, if $x \notin A$. If $\emptyset \neq C \subseteq A$

and C is connected, the connected component of A containing C, denoted by $\mathrm{cc}(A, C)$, is $\mathrm{cc}(X, x)$, with $x \in C$.

Definition 2.3. Let $A \subseteq \overline{\Omega}$. We call holes of A in $\overline{\Omega}$ the components of $\overline{\Omega} \setminus A$. Let $p_\infty \in \overline{\Omega}$ be a reference point. We call saturation of A with respect to p_∞, and note $\mathrm{Sat}(A, p_\infty)$, the set

$$\overline{\Omega} \setminus \mathrm{cc}(\overline{\Omega} \setminus A, p_\infty).$$

Let us immediately clarify that our notion of hole aims at the computation of saturated sets, which enable us to compute the tree of shapes of the image. Besides the case where $N = 2$, it does not coincide for $N \geq 3$ with the usual notion of differential topology, where a solid torus in $\mathbb{R}^3$ is said to have one hole. For us, a solid torus does not have a hole: its complement is made of a single connected component. And this is what matters to us in this work, which may be used in order to fuse the trees of connected components of upper and lower level sets in dimensions $N \geq 3$ (see Chap. 5 and Sect. 6.4.3). The case $N = 2$ is very special in our context and a more efficient algorithm will be described in Chaps. 6 and 7.

If $p_\infty \notin A$, let H be the hole of A containing p_∞. We have $\mathrm{Sat}(A, p_\infty) = \overline{\Omega} \setminus H$. We shall refer to H as the external hole of A and to other holes of A as its internal holes. If $p_\infty \in A$, we get $\mathrm{Sat}(A, p_\infty) = \overline{\Omega}$. Note that $\mathrm{Sat}(A, p_\infty)$ is the union of A and its internal holes.

The reference point p_∞ acts as a point at infinity. In all that follows, we assume that the point $p_\infty \in \overline{\Omega}$ on which saturations are based is fixed. To simplify our notation, when we write $\mathrm{Sat}(A)$, we mean implicitly $\mathrm{Sat}(A, p_\infty)$.

Figure 2.1 illustrates the above definition. On the upper left we can see the original set A (light gray) which is a subset of the whole space $\overline{\Omega}$ (dark gray). The next three images represent the saturations of A (other than $\overline{\Omega}$) with respect to its three holes.

We prove first some basic properties of the saturation.

Lemma 2.4. *Let $A \subseteq \overline{\Omega}$. If A is open (resp. closed) in $\overline{\Omega}$, then its saturated sets are open (resp. closed) in $\overline{\Omega}$.*

Proof. If A is open in $\overline{\Omega}$, $\overline{\Omega} \setminus \mathrm{Sat}(A)$, being a connected component of a closed set in $\overline{\Omega}$, is also closed in $\overline{\Omega}$. Thus $\mathrm{Sat}(A)$ is open in $\overline{\Omega}$. If A is closed in $\overline{\Omega}$, $\overline{\Omega} \setminus \mathrm{Sat}(A)$, being a connected component of an open set in $\overline{\Omega}$, is also open, as $\overline{\Omega}$ is locally connected. Thus $\mathrm{Sat}(A)$ is closed in $\overline{\Omega}$. □

Lemma 2.5. *The saturation* $\mathrm{Sat}(.)$, *as an operator acting on subsets of $\overline{\Omega}$, is*

(i) monotone: $A \subseteq B \Rightarrow \mathrm{Sat}(A) \subseteq \mathrm{Sat}(B)$;
(ii) idempotent: $\mathrm{Sat}\left[\mathrm{Sat}(A)\right] = \mathrm{Sat}(A)$.

Proof. The proofs are immediate and we shall skip them. □

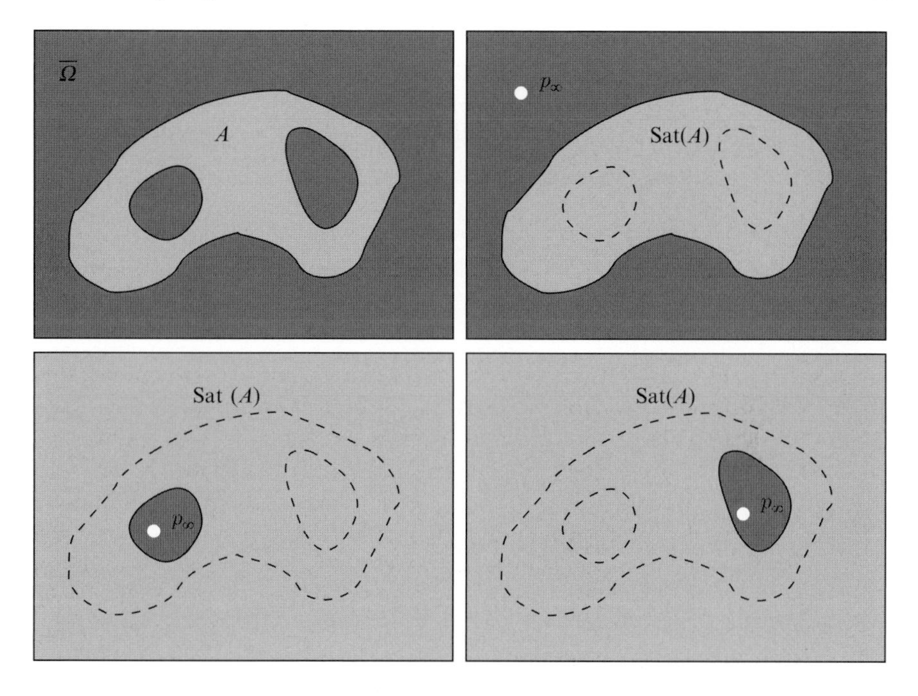

Fig. 2.1 A set A (in light gray, upper left image) and its saturations.

Lemma 2.6. *Let $A \subseteq \overline{\Omega}$ be a connected set and H be a hole of A. If H is an internal hole,* $\operatorname{Sat} H = H$*; if H is the external hole,* $\operatorname{Sat}(H) = \overline{\Omega}$.

Proof. If H is an internal hole, then $p_\infty \notin H$, and, therefore, $\operatorname{Sat}(H) \neq \overline{\Omega}$. Since H is a connected component of $\overline{\Omega} \setminus A$, the set $\overline{\Omega} \setminus H$ is connected ([84, IV.3, Theorem 3.3]). Hence $\overline{\Omega} \setminus H$ is a hole of H containing p_∞, and we conclude that $\operatorname{Sat}(H) = H$. If H is the external hole, then $p_\infty \in H$, and we have that $\operatorname{Sat}(H) = \overline{\Omega}$. □

We now prove three lemmas that will be useful in the sequel.

Lemma 2.7. *If $A \subseteq \overline{\Omega}$ is connected, then* $\operatorname{Sat}(A)$ *is also connected. Moreover, for any $H \in \mathcal{CC}(\overline{\Omega} \setminus A)$, ∂H is connected.*

Proof. If $\operatorname{Sat}(A) = \overline{\Omega}$, there is nothing to prove. Otherwise, $\overline{\Omega} \setminus \operatorname{Sat} A$ is a connected component of the complement of a connected set A, in a connected space $\overline{\Omega}$. Thanks to [51, §41,III,5], its complement, i.e., $\operatorname{Sat}(A)$, is connected.

To prove the second assertion, let $H \in \mathcal{CC}(\overline{\Omega} \setminus A)$. Since $\partial H = \overline{H} \cap \overline{\overline{\Omega} \setminus H}$, both sets $\overline{H}$ and $\overline{\overline{\Omega} \setminus H}$ are closed and connected, and $\overline{\Omega}$ is unicoherent, we deduce that ∂H is connected. □

Lemma 2.8. *Let X be an open (closed) connected subset of $\overline{\Omega}$. Then ∂* $\operatorname{Sat}(X)$ *is connected.*

Proof. If X is open, then $\operatorname{Sat}(X)$ is also an open set. Then $\overline{\operatorname{Sat}(X)}$ and $\overline{\Omega}\setminus\operatorname{Sat}(X)$ are closed connected sets. Then $\partial\operatorname{Sat}(X) = \overline{\operatorname{Sat}(X)}\cap(\overline{\Omega}\setminus\operatorname{Sat}(X))$ is also connected, since $\overline{\Omega}$ is unicoherent.

If X is closed, then $\operatorname{Sat}(X)$ and $\overline{\overline{\Omega}\setminus\operatorname{Sat}(X)}$ are both closed connected sets. Since $\overline{\Omega}$ is unicoherent, we conclude that $\partial\operatorname{Sat}(X) = \overline{\operatorname{Sat}(X)}\cap(\overline{\overline{\Omega}\setminus\operatorname{Sat}(X)}) = \operatorname{Sat}(X)\cap\overline{\overline{\Omega}\setminus\operatorname{Sat}(X)}$ is also connected. □

Lemma 2.9. *Let $A \subseteq \overline{\Omega}$. If H is a hole of A, then*

$$\partial H \subseteq \partial A.$$

As a consequence,

$$\partial\operatorname{Sat}(A) \subseteq \partial A.$$

Proof. H is a connected component of $\overline{\Omega}\setminus A$, so $\partial H \subseteq \partial(\overline{\Omega}\setminus A)$ (see [51, §44,III,3]), proving that $\partial H \subseteq \partial A$.

Let H be the external hole of A. Then

$$\partial\operatorname{Sat}(A) = \partial(\overline{\Omega}\setminus\operatorname{Sat}(A)) = \partial H \subseteq \partial A.$$

□

Lemma 2.10. *Let $A \subseteq \overline{\Omega}$ be such that* $\operatorname{Sat}(A) \neq \overline{\Omega}$. *Then* $\operatorname{Sat}(A) \subseteq \operatorname{Sat}(\partial A)$, *and, if A is closed, we get* $\operatorname{Sat}(A) = \operatorname{Sat}(\partial A)$.

Proof. Suppose first that A is closed. Since $\partial A \subseteq A$, taking the saturation of each member, we have that $\operatorname{Sat}(\partial A) \subseteq \operatorname{Sat}(A)$. To prove the other inclusion, let C be a connected component of $\overset{\circ}{A}$. It is easy to see that C, as any open connected set, is a hole of ∂C. Since $\partial C \subseteq \partial\overset{\circ}{A}$ and $C\cap\partial\overset{\circ}{A}=\emptyset$, we have that C is open and closed in $\overline{\Omega}\setminus\partial\overset{\circ}{A}$. Since C is also connected, then C is a hole of $\partial\overset{\circ}{A}$, and, therefore, $C \subseteq \operatorname{Sat}(\partial\overset{\circ}{A})$. Now, using that $\partial\overset{\circ}{A}\subseteq\partial A$, we conclude that $C \subseteq \operatorname{Sat}(\partial A)$. This proves that $\overset{\circ}{A}\subseteq\operatorname{Sat}(\partial A)$, and, thus,

$$A = \overset{\circ}{A}\cup\partial A \subseteq \operatorname{Sat}(\partial A).$$

Finally, taking saturations, we obtain $\operatorname{Sat}(A) \subseteq \operatorname{Sat}(\partial A)$.

For an arbitrary set A, $A \subseteq \overline{A}$ implies that $\operatorname{Sat}(A) \subseteq \operatorname{Sat}(\overline{A})$. If $\operatorname{Sat}(\overline{A}) \neq \overline{\Omega}$, the first part of the proof applies to $\overline{A}$ and we obtain that $\operatorname{Sat}(A) \subseteq \operatorname{Sat}(\overline{A}) = \operatorname{Sat}(\partial\overline{A})$. Finally, from $\partial\overline{A} \subseteq \partial A$, we conclude that $\operatorname{Sat}(A) \subseteq \operatorname{Sat}(\partial A)$.

Now, the remaining case is $\operatorname{Sat}(\overline{A}) = \overline{\Omega}$. If, in addition, we have that $\operatorname{Sat}(\partial A) = \overline{\Omega}$, the result is obvious. Otherwise, let $E = \overline{\Omega}\setminus\operatorname{Sat}(\partial A) \neq \emptyset$. Then E is a hole of ∂A, hence, it cannot meet both A *and* $\overline{\Omega}\setminus A$. If $E \subseteq A$, then $\operatorname{Sat} E \subseteq \operatorname{Sat} A \neq \overline{\Omega}$, so that $\operatorname{Sat}(E) = \emptyset$, a contradiction, since $E \subseteq \operatorname{Sat}(E)$.

Thus, E is contained in a hole H of A, H being not an internal hole, since, otherwise, $\text{Sat}(E) \subseteq \text{Sat}(H) \subseteq \text{Sat}(A)$, which is again impossible. Therefore, H is the external hole of A, and $E \subseteq H$ amounts to

$$\overline{\Omega} \setminus \text{Sat}(\partial A) \subseteq \overline{\Omega} \setminus \text{Sat}(A),$$

which proves the expected result. □

Obviously, the identity $\text{Sat}(A) = \text{Sat}(\partial A)$ in Lemma 2.10 does not hold if A is not closed. Indeed, if A is an open ball, then $\text{Sat}(A) = A$ while $\text{Sat}(\partial A) = \overline{A}$.

Lemma 2.11. *Let $A, B \subseteq \overline{\Omega}$ be connected sets such that $A \cap B = \emptyset$.*

(i) Then either $\text{Sat}(A) \subseteq \text{Sat}(B)$, *or* $\text{Sat}(B) \subseteq \text{Sat}(A)$, *or* $\text{Sat}(A) \cap \text{Sat}(B) = \emptyset$.
(ii) If, in addition, we assume that A, B are closed sets, then either $\text{Sat}(A) \subseteq int(\text{Sat}(B))$, *or* $\text{Sat}(B) \subseteq int(\text{Sat}(A))$, *or* $\text{Sat}(A) \cap \text{Sat}(B) = \emptyset$.

Proof. (*i*) Let $p \in \overline{\Omega}$ be the point with respect to which we take saturations. If $p \in \overline{A}$, then $\text{Sat}(A) = \overline{\Omega}$ and the result is true. Similarly, if $p \in B$, then $\text{Sat}(B) = \overline{\Omega}$ and again the Lemma holds. Thus, we may assume that $p \notin A \cup B$, hence A and B have each an external hole. Then, being connected, A is included in a hole H of B. If H is an internal hole, we get

$$\text{Sat}(A) \subseteq \text{Sat}(H) = H \subseteq satB.$$

If H is the external hole of B, then B is also included in a hole H' of A. If H' is an internal hole of A, the same proof above applies and we get $\text{Sat}(B) \subseteq \text{Sat}(A)$. If H' is the external hole of A, no internal hole of A meets B, and, thus, $\text{Sat}(A)$ is in the exterior hole of B. Therefore, no internal hole of B meets $\text{Sat}(A)$, implying that $\text{Sat}(A) \cap \text{Sat}(B) = \emptyset$.
(*ii*) If $\text{Sat}(A) \cap \text{Sat}(B) \neq \emptyset$, then by Lemma 2.11, both sets are nested. In case $\text{Sat}(A) \subseteq \text{Sat}(B)$, then A is contained in a hole of B. Since the holes of B are open sets, we have that $\text{Sat}(A) \subseteq int(\text{Sat}(B))$. Similarly, if $\text{Sat}(B) \subseteq \text{Sat}(A)$, then $\text{Sat}(B) \subseteq int(\text{Sat}(A))$. □

This lemma is illustrated in Fig. 2.2. From left to right and up to down we have the original sets A and B, the saturation of A and B with respect to a point p which is external to A (observe that this case gives $\text{Sat}(B) \subseteq \text{Sat}(A)$) and finally the saturation of both sets when p is external to B but internal to A (which corresponds to $\text{Sat}(A) \cap \text{Sat}(B) = \emptyset$).

Recall that a continuum in a topological space is a compact connected set and a domain is an open connected set.

Lemma 2.12. *Assume that $X \subseteq \overline{\Omega}$ is open or closed. Let $C \in \mathcal{CC}(X)$, and $x \in \text{Sat}(C) \setminus C$. Then there exists $O \in \mathcal{CC}(\overline{\Omega} \setminus X)$ such that $x \in \text{Sat}(O) \subseteq$*

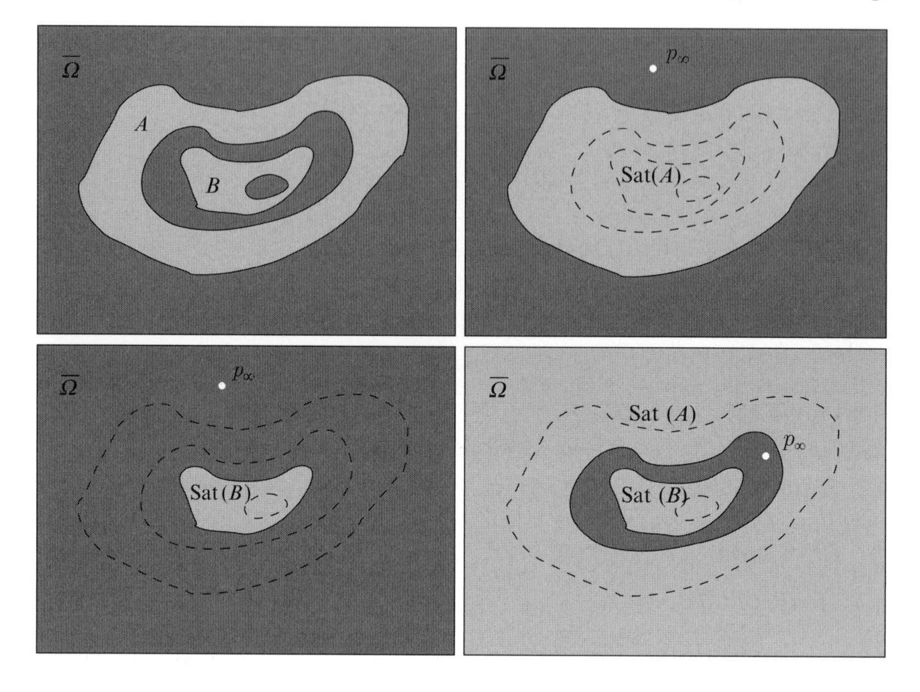

Fig. 2.2 Disjoint connected sets A and B and their saturations with respect to a point external to both sets or only external to B.

Sat(C). *Moreover, if X is open and Y is an internal hole of C, then there exists $O \in \mathcal{CC}(\overline{\Omega} \setminus X)$ such that $Y = \text{Sat}(O)$. The same statement holds if X is closed and has a finite number of connected components.*

Proof. Let us consider first the case where X is closed. Let H be the hole of C containing x. Observe that H is an open set whose boundary is connected, since $\overline{\Omega}$ is unicoherent.

Assume that

$$x \notin \text{Sat}(cc(\overline{\Omega} \setminus X, z)) \quad \text{for any } z \in H. \tag{2.1}$$

Notice that this implies that $x \in X$.

Consider the family of sets $cc(\overline{\Omega}\setminus X, z)$ where $z \in H \cap (\overline{\Omega}\setminus X)$. These sets are open and disjoint, thus, there are at most a countable number of them. The same property holds for their saturations, which we enumerate $C_1, C_2, \ldots$. Let us prove that for each n the set $D_n = \overline{H} \setminus \cup_{i=1}^n C_i$ is connected. Thanks to Lemma 2.2, we have $\overline{\Omega} \setminus \cup_{i=1}^n C_i$ connected and closed, so as $\overline{H}$, and their union is $\overline{\Omega}$. Therefore their intersection D_n is connected.

If the number of C_i is finite, this proves that $\overline{H} \cap X$ is connected. If this number is infinite, since D_n is a decreasing sequence of continua, then their intersection $\cap_n D_n = \overline{H} \cap X$ is also connected [51]. In any case, since $\partial H \subseteq C$, we deduce that $\overline{H} \cap X$ is included in C, a contradiction since $x \in \overline{H} \cap X$, but $x \notin C$. We conclude that (2.1) does not hold, and the first part of the Lemma is proved when X is closed.

Let S be the (finite) set of elements of $\mathcal{CC}(X)$ contained in Y, internal hole of C. Since $\cup_{T \in S} T$ is closed, $\tilde{Y} = \overline{\Omega} \setminus \bigcup_{T \in S} \mathrm{Sat}(T)$ is open and connected. Since $Y \cup \tilde{Y} = \overline{\Omega}$, we have that $O := Y \setminus \bigcup_{T \in S} \mathrm{Sat}(T)$ is connected. Then obviously $O \in \mathcal{CC}(\overline{\Omega} \setminus X)$. For any $T \in S$, $\mathrm{Sat}(T) \subseteq \mathrm{Sat}(O)$, so that $\cup_{T \in S} T \subseteq \mathrm{Sat}(O)$ and we get $Y = O \cup \cup_{T \in S} T \subseteq \mathrm{Sat}(O)$. Since $\mathrm{Sat}(O) \subseteq \mathrm{Sat}(Y) = Y$, we have that $Y = \mathrm{Sat}(O)$.

Now, assume that X is an open set, $C \in \mathcal{CC}(X)$, and Y is an internal hole of C. Since $\partial Y \subseteq \overline{\Omega} \setminus X$ and ∂Y is connected, we define $O = cc(\overline{\Omega} \setminus X, \partial Y)$. Since $\partial Y \subseteq O \subseteq Y$, using the monotonicity of Sat we have:

$$\mathrm{Sat}\, \partial Y \subseteq \mathrm{Sat}(O) \subseteq \mathrm{Sat}(Y).$$

But as Y is closed, $\mathrm{Sat}(\partial Y) = \mathrm{Sat}(Y)$ and as Y is a hole, $\mathrm{Sat}(Y) = Y$, so that we get $Y = \mathrm{Sat}(O)$. □

Lemma 2.13. *(i) Let* $(K_n)_{n \in \mathbb{N}}$ *be a decreasing sequence of continua,* $K = \cap_n K_n$. *Then* $\mathrm{Sat}(K) = \cap_n \mathrm{Sat}(K_n)$.
(ii) Let $(O_n)_{n \in \mathbb{N}}$ *be an increasing sequence of domains,* $O = \cup_n O_n$. *Then* $\mathrm{Sat}(O) = \cup_n \mathrm{Sat}(O_n)$.

Proof. (*i*) If $p_\infty \in K \subseteq K_n$, then $\mathrm{Sat}(K) = \mathrm{Sat}(K_n)$ for all n and the Lemma is obviously true. Thus, we may assume that $p_\infty \notin K$. Then, for n large enough, $p_\infty \notin K_n$. By the monotonicity of the saturation, $\mathrm{Sat}(K) \subseteq \cap_n \mathrm{Sat}(K_n)$. Let $q \notin \mathrm{Sat}(K)$, $q \neq p_\infty$, and let γ be an arc joining p_∞ to q such that $Im(\gamma) \cap \mathrm{Sat}(K) = \emptyset$. Then, for n large enough, we have $Im(\gamma) \cap K_n = \emptyset$. Thus, both p_∞ and q are in the same hole of K_n. Hence, $Im(\gamma) \cap \mathrm{Sat}(K_n) = \emptyset$, and, in particular, $q \notin \mathrm{Sat}(K_n)$, for n large enough. Thus, $q \notin \cap_n \mathrm{Sat}(K_n)$. We conclude that $\cap_n \mathrm{Sat}(K_n) \subseteq \mathrm{Sat}(K)$.
(*ii*) Obviously, the inclusion $\cup_n \mathrm{Sat}(O_n) \subseteq \mathrm{Sat}(O)$ holds. If $p_\infty \in O$, also $p_\infty \in O_n$ for n large enough and the identity holds. Thus, we may assume that $p_\infty \notin O$. Let $q \notin \mathrm{Sat}(O_n)$ for all n. Then $q \in H_n$ where H_n is the external hole of $\mathrm{Sat}(O_n)$ which also contains p_∞. Since H_n is a continuum, also is $\cap_n H_n$ and it contains both q and p_∞. Then

$$\begin{aligned} \cap_n H_n \subseteq \cap_n(\overline{\Omega} \setminus O_n) &= \overline{\Omega} \setminus \cup_n O_n \\ &= \overline{\Omega} \setminus O. \end{aligned}$$

Thus $\cap_n H_n$ is in a hole of O which contains both q and p_∞. We conclude that $q \notin \mathrm{Sat}(O)$. The identity $\cup_n \mathrm{Sat}(O_n) = \mathrm{Sat}(O)$ holds. □

2.3 The Tree of Shapes

The concept of tree can be defined in any ordered set. Since the general definition will not be necessary in this book, we restrict our attention to trees formed by subsets of $\overline{\Omega}$. As usual, we denote by $\mathcal{P}(\overline{\Omega})$ the Boolean algebra of all subsets of $\overline{\Omega}$.

Definition 2.14. Let $\mathcal{T} \subseteq \mathcal{P}(\overline{\Omega})$. We say that $\mathcal{T}$ is a tree of subsets of $\overline{\Omega}$ if

(i) $\mathcal{T}$ contains $\overline{\Omega}$,
(ii) If $C, D \in \mathcal{T}$, then either $C \cap D \neq \emptyset$, $C \subseteq D$ or $D \subseteq C$. In the last two cases we shall say that C and D are nested.

The elements of the tree will be called nodes. Observe that, thanks to *(ii)*, there is no cycle in a tree: if A, B_1, B_2, C are nodes and $A \subseteq B_i \subseteq C$, $i = 1, 2$, then $B_1 \cap B_2 \neq \emptyset$, hence the sets B_1 and B_2 must be nested.

An image u is a map from $\overline{\Omega}$ to $\mathbb{R}$, supposed to be *upper semicontinuous.* The lower and upper level sets of u are the sets

$$[u < \lambda] = \left\{x \in \overline{\Omega},\, u(x) < \lambda\right\} \qquad [u \geq \lambda] = \left\{x \in \overline{\Omega},\, u(x) \geq \lambda\right\}.$$

Lower level sets are open, while upper level sets are closed. The asymmetry comes from the assumption that u is *upper* semicontinuous, and would be inverted if we assumed u to be *lower* semicontinuous.

Upper (resp. lower) level sets are an equivalent description of the image, since we can reconstruct it by the formula

$$u(x) := \sup\{\lambda \in \mathbb{R} : x \in [u \geq \lambda]\} \qquad (\text{resp. } u(x) := \inf\{\lambda \in \mathbb{R} : x \in [u < \lambda]\}).$$

Let us denote

$$\mathcal{U}(u) = \{X :\, X \in \mathcal{CC}([u \geq \lambda]),\, \lambda \in \mathbb{R}\}$$

and

$$\mathcal{L}(u) = \{X :\, X \in \mathcal{CC}([u < \lambda]),\, \lambda \in \mathbb{R}\}.$$

Observe that if $\lambda \geq \mu$ and $X \in \mathcal{CC}([u \geq \lambda])$, $Y \in \mathcal{CC}([u \geq \mu])$, then either $X \cap Y = \emptyset$ or $X \subseteq Y$. Similarly, if $X \in \mathcal{CC}([u < \lambda])$, $Y \in \mathcal{CC}([u < \mu])$, then either $X \cap Y = \emptyset$ or $Y \subseteq X$. That is, any two connected components of any two upper (resp. lower) level sets are either disjoint or nested. Hence both $\mathcal{U}(u)$ and $\mathcal{L}(u)$ are trees.

Definition 2.15. Given an image u, we call shapes of inferior (resp. superior) type the sets

$$\mathrm{Sat}(\mathrm{cc}([u < \mu], x)) \quad (\text{resp. } \mathrm{Sat}(\mathrm{cc}([u \geq \lambda], x))),$$

where $\lambda, \mu \in \mathbb{R}$, $x \in \overline{\Omega}$. We call shape of u any shape of inferior or superior type. We denote by $\mathcal{S}(u)$ the shapes of u.

Having fixed the point p_∞ on which the saturation is based, the shapes of the image depend on p_∞. If we choose $p_\infty \in \partial\Omega$ and we assume that the function u is constant in a neighborhood of $\partial\Omega$, then we may consider that our notion of shape becomes intrinsic.

We note that, by definition, shapes of superior type are closed, while shapes of inferior type are open. Since shapes are connected, the only shapes of both types are $\emptyset$ and $\overline{\Omega}$.

The rest of the section is devoted to prove that shapes have a tree structure, and to define common features of $\mathcal{L}(u)$, $\mathcal{U}(u)$ and the tree of shapes, like branches and leaves.

2.3.1 Tree Structure of the Set of Shapes

Theorem 2.16. *Any two shapes are either disjoint or nested. Hence, $\mathcal{S}(u)$ is a tree.*

Proof. Let $A = \mathrm{Sat}(\tilde{A})$ and $B = \mathrm{Sat}(\tilde{B})$, $\tilde{A}$ and $\tilde{B}$ being connected components of level sets of u.

If $\tilde{A}$ and $\tilde{B}$ are of the same type, then they are either nested or disjoint. In the first case, A and B are nested because Sat is a monotone operator, while, in the second case, Lemma 2.11 permits to conclude. Thus, we may assume that $\tilde{A}$, $\tilde{B}$ are of different type and $\tilde{A} \cap \tilde{B} \neq \emptyset$. Let $x \in \tilde{A} \cap \tilde{B}$. By interchanging names, if necessary, we may assume that $\tilde{A} = \mathrm{cc}([u \geq \lambda])$ and $\tilde{B} = \mathrm{cc}([u < \mu])$. Observe that $\lambda \leq u(x) < \mu$, hence $\lambda < \mu$.

The set $\tilde{B}$, being a connected component of $[u < \mu]$, is open in $\overline{\Omega}$ and closed in $[u < \mu]$, hence,

$$[u < \mu] \cap \partial\tilde{B} = [u < \mu] \cap (\overline{\tilde{B}} \setminus \tilde{B}) = \emptyset.$$

This proves that $\partial\tilde{B} \subseteq [u \geq \mu] \subseteq [u \geq \lambda]$ and, according to Lemma 2.9, $\partial B \subseteq \partial\tilde{B} \subseteq [u \geq \lambda]$.

Since A is connected, if $\partial B \cap A = \emptyset$, we have that either $A \subseteq B$ or $A \subseteq \overline{\Omega} \setminus \overline{B}$, the last case being impossible because $A \cap B \neq \emptyset$. Thus, we may assume that $\partial B \cap A \neq \emptyset$. First, we observe that ∂B is connected. Indeed, if H is the external hole of B, the connectedness of ∂B follows from the identity

$$\partial B = \overline{B} \cap \overline{H},$$

and the unicoherency of $\overline{\Omega}$, since $\overline{B}$ and $\overline{H}$ are connected, being closures of connected sets.

From the above, we conclude that ∂B is contained in a connected component C of $[u \geq \lambda]$. If $C \cap \tilde{A} = \emptyset$, then C is in a hole of $\tilde{A}$, but not in the external one as $\partial B \cap A \neq \emptyset$. The other case being $C = \tilde{A}$, we conclude that, in any case, we have that $\partial B \subseteq A$. Now, using Lemma 2.10, we obtain

$$B = \mathrm{Sat}(B) \subseteq \mathrm{Sat}(\partial B) \subseteq \mathrm{Sat}(A) = A.$$

The root of the tree is

$$\overline{\Omega} = \mathrm{Sat}([u < 1 + \max_{\overline{\Omega}} u]).$$

□

2.3.2 Branches, Monotone Sections and Leaves

In this section $\mathcal{T}$ denotes a tree of subsets of Ω.

Definition 2.17. Let $A \subseteq B \subseteq \overline{\Omega}$. We define $[A, B]$ as the interval of $\mathcal{T}$ between A and B, i.e.,

$$[A, B] = \{S : S \in \mathcal{T},\ A \subseteq S \subseteq B\}.$$

We also define

$$\inf[A, B] = \bigcap_{S \in [A,B]} S \text{ and } \sup[A, B] = \bigcup_{S \in [A,B]} S.$$

We say that T is a limit node of $\mathcal{T}$ if it is the infimum or the supremum of a nonempty interval of $\mathcal{T}$.

In using the notation $[A, B]$ we implicitly understand that it is an interval of $\mathcal{T}$. In case that we are considering several trees at the same time, we may use the notation $[A, B]_{\mathcal{T}}$ to stress the fact that we refer to an interval of $\mathcal{T}$.

We call the intervals of shapes the intervals of $\mathcal{S}(u)$. The limit nodes of this tree will be called limit shapes.

Observe that, if $A \subseteq B \subseteq \overline{\Omega}$, then $[A, B] = [\inf[A, B], \sup[A, B]]$. Thus, when considering an interval, we may always assume that its extreme sets are limit nodes.

Definition 2.18. Let $B \subseteq \overline{\Omega}$. We say that B contains a bifurcation in $\mathcal{T}$ if there exist $S, T \in \mathcal{T}$ such that $S, T \subseteq B$ and $S \cap T = \emptyset$. Let $A \subseteq B \subseteq \overline{\Omega}$. We say that there is a bifurcation between A and B if $\inf[A, B] \neq \emptyset$ and there is $S \in \mathcal{T}$ such that $S \subseteq B$ and $S \cap \inf[A, B] = \emptyset$.

Definition 2.19. Let $A \subseteq B \subseteq \overline{\Omega}$. We say that $[A, B]$ is a branch of $\mathcal{T}$ if there is no bifurcation between A and B. If $\mathcal{T} = \mathcal{S}(u)$, we say that $[A, B]$ is a monotone section if $[A, B]$ is a branch with all shapes of the same type.

We say that a branch $[A, B]$ (resp., a monotone section) contains $x \in \overline{\Omega}$ if there is a node $S \in [A, B]$ such that $x \in S$.

Definition 2.20. Let $x \in \overline{\Omega}$. We define S_x as the smallest limit node containing x, i.e.,

$$S_x = \inf[\{x\}, \overline{\Omega}] = \bigcap_{S \in \mathcal{T}, x \in S} S.$$

The upper branch at x is the set

$$\mathcal{B}_x = \{S : \ S \in \mathcal{T}, \ x \in S, \ \text{and } [S_x, S] \text{ is a branch.}\}$$

Definition 2.21. We call leaf of $\mathcal{T}$, or simply, a leaf, any limit node $L = \inf[A, B]$ containing no other shape.

Proposition 2.22. *Any two limit nodes of $\mathcal{T}$ are either nested or disjoint.*

Proof. Assume first that T is a node and S is a limit node of $\mathcal{T}$ such that $T \cap S \neq \emptyset$. Suppose that $S = \inf[A, B]$. Then $T \cap R \neq \emptyset$ for all $R \in [A, B]$. Since $\mathcal{T}$ is a tree, for any $R \in [A, B]$, either $T \subseteq R$ or $R \subseteq T$. Then, either there is $R \in [A, B]$ such that $R \subseteq T$, or $T \subseteq R$, for all $R \in [A, B]$. In the first case, we have that $S = \inf[A, B] \subseteq T$. In the second one, $T \subseteq \inf[A, B] = S$.

Suppose that $S = \sup[A, B]$. Since $T \cap S \neq \emptyset$, there is $R \in [A, B]$ such that $T \cap R \neq \emptyset$. Observe that $S = \sup[R, B]$ and $T \cap Q \neq \emptyset$ for all $Q \in [R, B]$. Again, since $\mathcal{T}$ is a tree, for any $Q \in [R, B]$, either $T \subseteq Q$ or $Q \subseteq T$. Then, either there is $Q \in [R, B]$ such that $T \subseteq Q$, or $Q \subseteq T$ for all $Q \in [R, B]$. In the first case, we have that $T \subseteq \sup[R, B] = S$. In the second one, $S = \sup[A, B] \subseteq T$.

Now, we suppose that both S and T are limit nodes. First, assume that $S = \inf[A, B]$. If $S \cap T \neq \emptyset$, then $R \cap T \neq \emptyset$ for all $R \in [A, B]$. Then, either there is some $R \in [A, B]$ such that $R \subseteq T$, or $T \subseteq R$ for all $R \in [A, B]$. In the first case, $S \subseteq T$, in the second, $T \subseteq S$. Now, assume that $S = \sup[A, B]$. If $S \cap T \neq \emptyset$, then there is some $R \in [A, B]$ such that $R \cap T \neq \emptyset$. Observe that $S = \sup[R, B]$ and $T \cap Q \neq \emptyset$ for all $Q \in [R, B]$. By the first part of the proof, for any $Q \in [R, B]$, either $T \subseteq Q$ or $Q \subseteq T$. Then, either there is $Q \in [R, B]$ such that $T \subseteq Q$, or $Q \subseteq T$, for all $Q \in [R, B]$. In the first case, we have that $T \subseteq \sup[R, B] = S$, in the second, $S = \sup[A, B] \subseteq T$. □

For the tree of shapes we have the following complementary information.

Proposition 2.23. *Any limit shape can be expressed as a countable union or intersection of shapes of the same type.*

Proof. Let S be a limit shape. Assume that $S = \sup[A, B]$. Define

$$[A, B]_+ := \{T \in [A, B] : T \text{ is of upper type}\},$$

and

$$[A, B]_- := \{T \in [A, B] : T \text{ is of lower type}\}.$$

Since $\sup[A, B] = \sup[A, B]_- \cup \sup[A, B]_+$, and the set is linearly ordered, we have that either $\sup[A, B] = \sup[A, B]_+$, or $\sup[A, B] = \sup[A, B]_-$.

Assume that $S = \sup[A, B]_+ = \bigcup_{S \in [A,B]_+} S$. Since the elements of $[A, B]_+$ are closed sets, by [51, §19,VII,1] we know that the we may establish an order isomorphism between the shapes of $[A, B]_+$ and an interval J of $[0, 1]$. Since there is a countable set $J' \subseteq J$ such that $\sup J = \sup J'$, we conclude that there is a countable set $\{S_n\}_n \subseteq [A, B]_+$ such that $\sup[A, B]_+ = \cup_n S_n$.

Assume that $S = \sup[A, B]_- = \bigcup_{S \in [A,B]_-} S$. By Lindelöf's theorem [51, §17,I,1], there is a countable family $\{S_n\}_n \subseteq [A, B]_-$ such that $\sup[A, B]_- = \cup_n S_n$.

The case where $S = \inf[A, B]$ can be proved in an analogous way. □

Proposition 2.24. *Let $A_1, A_2 \neq \emptyset$, and $[A_1, B_1]$, $[A_2, B_2]$ two branches of $\mathcal{T}$ such that*

$$[A_1, B_1] \cap [A_2, B_2] \neq \emptyset.$$

Without loss of generality we may assume that A_1, A_2, B_1, B_2 are limit nodes. Then $[A_1 \cap A_2, B_1 \cup B_2]$ is a branch.

Proof. Let $S \in \mathcal{T}$ be such that $A_1 \subseteq S \subseteq B_1$ and $A_2 \subseteq S \subseteq B_2$. Thus $B_1 \cap B_2 \neq \emptyset$, and, therefore, B_1 and B_2 are nested. The limit nodes A_1 and A_2 are also nested. Indeed, we observe that $A_2 = \inf[A_2, S]$. For any $R \in [A_2, S]$, $R \subseteq B_1$ and since there is no bifurcation in $[A_1, B_1]$, $A_1 \cap R \neq \emptyset$. If there is $R \in [A_2, S]$ such that $R \subseteq A_1$, then $A_2 = \inf[A_2, S] \subseteq A_1$. If this does not happen, $A_1 \subseteq R$ for all $R \in [A_2, S]$, and we have that $A_1 \subseteq \inf[A_2, S] = A_2$.

By symmetry, we may assume that either

$$A_1 \subseteq B_1 \subseteq A_2 \subseteq B_2, \text{ or} \quad (i)$$
$$A_1 \subseteq A_2 \subseteq B_1 \subseteq B_2, \text{ or} \quad (ii)$$
$$A_1 \subseteq A_2 \subseteq B_2 \subseteq B_1. \quad (iii)$$

If (i) holds, then

$$A_1 \subseteq S \subseteq B_1 \subseteq A_2 \subseteq S \subseteq B_2,$$

and $B_1 = A_2 = S$. This case can be subsumed under the case (ii). If (iii) holds, then $[A_1 \cap A_2, B_1 \cup B_2] = [A_1, B_1]$ is a branch. Thus, we may assume that (ii) holds. We prove that there is no bifurcation in $[A_1, B_2]$.

Let R a node such that $R \subseteq B_2$. Since $[A_2, B_2]$ is a branch, $R \cap A_2 \neq \emptyset$. By Proposition 2.22, either $A_2 \subseteq R$ or $R \subseteq A_2$. In the first case, $A_1 \subseteq R$, while, in the second case, $R \subseteq B_1$ and, since $[A_1, B_1]$ is a branch, $R \cap A_1 \neq \emptyset$. This holds for any $R \subseteq B_2$, proving that $[A_1, B_2]$ is a branch. □

Proposition 2.24 permits us to define the maximal branch containing a given node $S \in \mathcal{T}$. Indeed, we define $\mathcal{B}_{\mathcal{T}}(S)$, the maximal branch containing S, as

$$\mathcal{B}_{\mathcal{T}}(S) = \cup\{[A,B] : [A,B] \text{ is a branch of } \mathcal{T} \text{ s.t. } S \in [A,B]\}. \tag{2.2}$$

When the tree $\mathcal{T}$ is implicitly understood, we simply denote $\mathcal{B}(S)$ instead of $\mathcal{B}_{\mathcal{T}}(S)$. We observe that

$$\mathcal{B}(S) = [\inf \mathcal{B}(S), \sup \mathcal{B}(S)].$$

As a consequence of Proposition 2.24 we obtain the following corollary.

Corollary 2.25. *Let $x \in \overline{\Omega}$. The upper branch at x, $\mathcal{B}_x$, is a branch.*

Proposition 2.26. *Any two distinct leaves in $\mathcal{T}$ are disjoint.*

Proof. Let L_1, L_2 be two distinct leaves. Assume that $L_1 = \inf[A_1, B_1]$, $L_2 = \inf[A_2, B_2]$. If they are not disjoint, then either $L_1 \subseteq L_2$ or $L_2 \subseteq L_1$. By symmetry, we may assume that $L_1 \subseteq L_2$. Then, for any $S \in [A_1, B_1]$ and any $T \in [A_2, B_2]$, we have that $S \cap T \subseteq L_1 \neq \emptyset$. Thus, S and T are nested. Thus either

$$\exists S \in [A_1, B_1] \text{ s.t. } \forall T \in [A_2, B_2],\ S \subseteq T, \text{ or} \tag{2.3}$$

$$\forall S \in [A_1, B_1],\ \exists T \in [A_2, B_2] \text{ s.t. } T \subseteq S. \tag{2.4}$$

If (2.3) holds, then $S \subseteq L_2$. If $S \neq L_2$, then L_2 would not be a leaf. Thus, $S = L_2$. Since $L_1 = \inf[A_1, S]$, if $R \in [A_1, S]$, $R \subseteq L_2$. If, for some $R \in [A_1, S]$, $R \neq L_2$, then L_2 would not be a leaf. Thus, for any $R \in [A_1, S]$, $R = L_2$. Therefore $L_1 = L_2$, a contradiction. If (2.4) holds, then $L_2 \subseteq L_1$. Hence, $L_1 = L_2$. This contradiction proves that L_1 and L_2 must be disjoint. □

Definition 2.27. Let $A \subseteq B \subseteq \overline{\Omega}$. We say that the interval $[A, B]$ of $\mathcal{S}(u)$ has a definite type if all its shapes are of the same type.

Proposition 2.28. *Let $A_1, A_2 \neq \emptyset$, and let $[A_1, B_1]$, $[A_2, B_2]$ be two monotone sections (of $\mathcal{S}(u)$) of the same type such that $[A_1, B_1] \cap [A_2, B_2] \neq \emptyset$. Without loss of generality we may assume that A_1, A_2, B_1, B_2 are limit shapes. Then $[A_1 \cap A_2, B_1 \cup B_2]$ is a monotone section.*

Proof. By Proposition 2.24, we know that $[A_1 \cap A_2, B_1 \cup B_2]$ is a branch. As in the proof of Proposition 2.24, we may reduce the study to one of the three cases (i), (ii) or (iii). In case (iii), the conclusion is immediate. As we observed above, (i) can be subsumed under case (ii). Thus, assume that (ii) holds. Assume that the type of both intervals is the superior one. Let $R \in [A_1, B_2]$. Since $R \cap A_2 \neq \emptyset$, then either $A_2 \subseteq R$, or $R \subseteq A_2$. In the first case, $R \in [A_2, B_2]$, in the second, $R \in [A_1, B_1]$. In any case, R is a shape of superior type. Thus $[A_1, B_2]$ is an interval with all shapes of superior type. The proof is similar if both intervals are of inferior type. □

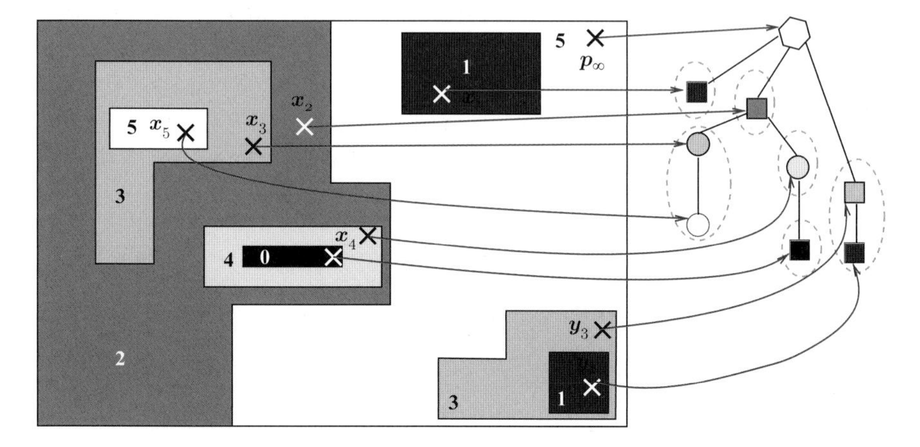

Fig. 2.3 Example of tree associated to an image. Left: the sample image. Right: Associated tree. Squares represent shapes of inferior type, while circles represent shapes of superior type. The maximal monotone sections are indicated by ellipses. Various points x are marked in the image, and, for each of them, a line points to the smallest limit shape S_x associated to it. A red border denotes a leaf.

Propositions 2.28 permit us to define the maximal monotone section $\mathcal{M}(S)$ containing a given shape $S \in \mathcal{S}(u)$ as

$$\mathcal{M}(S) = \cup\{[A, B] : [A, B] \text{ is a monotone section s.t. } S \in [A, B]\}. \quad (2.5)$$

We observe that

$$\mathcal{M}(S) = [\inf \mathcal{M}(S), \sup \mathcal{M}(S)].$$

These various notions are illustrated for the tree of shapes in Fig. 2.3.

2.4 Reconstruction of the Image From Its Tree

This section explains how an image can be reconstructed from its tree; more precisely, we prove that the image is perfectly defined by its shapes and the levels at which they are extracted.

Definition 2.29. We call lower, resp. upper, level shape at level $\lambda \in \mathbb{R}$ at $x \in \overline{\Omega}$ a pair $(\lambda, S = \mathrm{Sat}(C))$ with $x \in S$ and $C \in \mathcal{CC}([u < \lambda])$, resp. $C \in \mathcal{CC}([u \geq \lambda])$. We note LS_x the set of all lower and upper level shapes at x for any λ. We also note $G_{\lambda,x} = \{S : (\lambda, S) \text{ is a lower shape}\}$, $F_{\lambda,x} = \{S : (\lambda, S) \text{ is an upper shape}\}$ and $\mathrm{LS}_{\lambda,x} = G_{\lambda,x} \cup F_{\lambda,x}$.

The relation $(\lambda, S) \preccurlyeq (\mu, T)$, defined as $S \subsetneq T$ or ($S = T$ and $\lambda \leq \mu$, if $S \in G_{\lambda,x}, T \in G_{\mu,x}$; resp. $\lambda \geq \mu$, if $S \in F_{\lambda,x}, T \in F_{\mu,x}$), is a total order relation in LS_x, provided that $S \neq \overline{\Omega} \neq T$.

Lemma 2.30. *Let $x \in \overline{\Omega}$, A be a connected component of a level set of u, and $(\lambda, \mathrm{Sat}(A)) \in \mathrm{LS}_x$. If $x \notin A$, then there exists $(\lambda, \mathrm{Sat}(B)) \in \mathrm{LS}_x$ such that $\mathrm{Sat}(B) \subsetneq \mathrm{Sat}(A)$ and B is of a different type than A.*

Proof. Let T be the hole of A containing x. If $A = \mathrm{cc}([u < \lambda])$, then $\partial T \subseteq [u \geq \lambda]$, and, since ∂T is connected, we have that $(\lambda, \mathrm{Sat}(T) = T) \in F_{\lambda,x}$, with $T \subsetneq \mathrm{Sat}(A)$. Thus, we may assume that $A = \mathrm{cc}([u \geq \lambda])$.

Let H be the family of connected components of $[u < \lambda]$ inside T. Since H is a family of open and disjoint sets, we can enumerate its elements, H_1, $\ldots$, $H_n \ldots$ Let $D_n = \overline{T} \setminus \bigcup_{i=1}^{n} \mathrm{Sat}(H_i)$. Clearly, D_n is closed. Let us prove that D_n is connected. Let $G \neq \emptyset$ be an open and closed subset of D_n. Now, it is easy to check that

$$\partial D_n = \partial T \cup \bigcup_{i=1}^{n} \partial \,\mathrm{Sat}(H_i),$$

and we observe that each term of this union is connected. Hence, if G meets one of $\partial \,\mathrm{Sat}(H_i)$, it contains it. Let G' be the union of G and the sets $\mathrm{Sat}(H_i)$ whose boundary is contained in G. Obviously G' is open in $\overline{T}$. We observe that, if the boundary of $\mathrm{Sat}(H_i)$ is contained in G, then $G \cup \mathrm{Sat}(H_i) = \overline{G \cup \mathrm{Sat}(H_i)}$, and, therefore, G' is also closed in $\overline{T}$. Due to the connectedness of $\overline{T}$, $G' = \overline{T}$ and, therefore, $G = D_n$. This proves that D_n is connected.

If the number of H_i is finite, this proves that $\overline{T} \setminus \bigcup_i \mathrm{Sat}(H_i)$ is connected. If there are infinitely many, the sets D_n form a decreasing sequence of continua, thus, their intersection is a continuum, thanks to Zoretti's theorem ([51, §42,II,4]).

Since $\partial T \subseteq A$ and is connected, we have that $\overline{T} \setminus \bigcup_i \mathrm{Sat}(H_i) \subseteq A$. Now, since $x \notin A$, it follows that $x \in \bigcup_i \mathrm{Sat}(H_i)$, which implies the conclusion of the Lemma. □

2.4.1 Direct Reconstruction

Theorem 2.31. *Let $x \in \overline{\Omega}$. If $S_x \neq \overline{\Omega}$, then*

$$u(x) = \lim_{\mathrm{LS}_x \ni (\lambda, S) \searrow} \lambda. \tag{2.6}$$

If $S_x = \overline{\Omega}$, then

$$u(x) = \inf\{\lambda : \overline{\Omega} \in G_{\lambda,x}\} = \max\{\lambda : \overline{\Omega} \in F_{\lambda,x}\} \tag{2.7}$$

Equation (2.6) deserves some explanation. It must be interpreted as

$$\forall \varepsilon > 0, \exists (\lambda, S) \in \mathrm{LS}_x, \forall (\mu, T) \preccurlyeq (\lambda, S), |u(x) - \mu| \leq \varepsilon.$$

Loosely speaking, $u(x)$ is the level of the smallest shape containing x. But this "smallest shape" does not necessarily exist (it is actually a limit shape) and the relation $\preccurlyeq$ takes into account that the same shape can sometimes be extracted at several levels.

Proof. Suppose first that $S_x = \overline{\Omega}$. Let $\lambda \leq u(x)$ and $C_1 = \mathrm{cc}([u \geq \lambda], x)$. Then $\mathrm{Sat}\, C_1 = \overline{\Omega}$. Now, if $C_2 = \mathrm{cc}([u < \lambda])$, is such that $C_2 \cap C_1 = \emptyset$, we have that $\mathrm{Sat}(C_2) \neq \overline{\Omega}$, and, therefore, $\mathrm{Sat}\, C_2$ does not contain x. This proves that

$$u(x) \leq \inf\{\lambda : \overline{\Omega} \in G_{\lambda,x}\}.$$

Now, if $\lambda_n = u(x) + 1/n$, it is clear that $\overline{\Omega} \in G_{\lambda_n,x}$, thus,

$$\inf\{\lambda : \overline{\Omega} \in G_{\lambda,x}\} \leq \lambda_n$$

which gives the equality with $u(x)$ as n tends to infinity. The second equality follows by using the same argument, the maximum being reached this time since $\lambda > u(x) \Rightarrow \overline{\Omega} \notin F_{\lambda,x}$.

We assume now that $S_x \subsetneq \overline{\Omega}$. There is some shape $S^* \subsetneq \overline{\Omega}$, containing S_x, and let λ^* be its associated level. Let us consider the shape $S_0 = \mathrm{Sat}(\mathrm{cc}([u \geq u(x)], x))$. It is clear that $(u(x), S_0) \in \mathrm{LS}_x$. Let $(\lambda, S) \in \mathrm{LS}_x$ be such that $(\lambda, S) \preccurlyeq (u(x), S_0)$ if $S_0 \subsetneq \overline{\Omega}$, or $S \subsetneq \overline{\Omega}$ if $S_0 = \overline{\Omega}$. We claim that $\lambda \geq u(x)$, this inequality being strict if $S_0 = \overline{\Omega}$.

If S is a lower shape, we have that $S \subsetneq S_0$. Since $S = \mathrm{Sat}(\mathrm{cc}([u < \lambda]))$, if we have that $\lambda \leq u(x)$, we deduce that this $\mathrm{cc}([u < \lambda])$ does not contain x, and, thus, that x is in one of its internal holes, namely in $\mathrm{cc}([u \geq u(x)], x)$ and, therefore, also its saturation, which is impossible. Thus, we must have that $\lambda > u(x)$.

If S is an upper shape, we write $S = \mathrm{Sat}(\mathrm{cc}([u \geq \lambda]))$. If $x \in \mathrm{cc}[u \geq \lambda]$, we have that $u(x) \geq \lambda$, in which case $\mathrm{cc}([u \geq \lambda]) \supseteq \mathrm{cc}([u \geq u(x)], x)$. It follows that $S_0 \neq \overline{\Omega}$ and $\mathrm{cc}([u \geq \lambda] = \mathrm{cc}([u \geq u(x)], x)$. Then the order relation yields that $\lambda \geq u(x)$, and, thus, we have the equality. Thus, we may assume that x is in an internal hole of $\mathrm{cc}([u \geq \lambda])$. Since $\mathrm{cc}([u \geq u(x), x)$ is not included in this hole, it must intersect $\mathrm{cc}([u \geq \lambda])$, and, hence, it contains it (strictly). This implies that $\lambda > u(x)$.

For $\varepsilon > 0$, let us consider the shape $S_\varepsilon = \mathrm{Sat}(\mathrm{cc}([u < u(x) + \varepsilon], x))$ and assume that $S_\varepsilon \neq \overline{\Omega}$. Let $(\lambda, S) \in \mathrm{LS}_x$ be such that $(\lambda, S) \preccurlyeq (u(x) + \varepsilon, S_\varepsilon)$.

Suppose that S is an upper shape, $S = \mathrm{cc}([u \geq \lambda]) \neq \overline{\Omega}$. If $\lambda \geq u(x) + \varepsilon$, then $x \notin \mathrm{cc}([u \geq \lambda])$ and there is an internal hole H of this set containing x. This implies that $\mathrm{cc}([u < u(x) + \varepsilon], x)$ is also contained in H, and since, by assumption, $S \neq \overline{\Omega}$, we have that $\mathrm{Sat}(H) = H$. It follows that $S_\varepsilon \subseteq H \subsetneq S$, which is contrary to the assumption that $S \subseteq S_\varepsilon$. Thus, $\lambda < u(x) + \varepsilon$.

If $S = \mathrm{Sat}(\mathrm{cc}([u < \lambda]))$ is a lower shape, and $S = S_\varepsilon$, we must have that $\lambda \leq u(x) + \varepsilon$. In case $S \subsetneq S_\varepsilon$, we have either $x \in \mathrm{cc}([u < \lambda])$, in which case $\lambda < u(x) + \varepsilon$, or x is in an internal hole H of $\mathrm{cc}([u < \lambda])$. In the second case, since $\mathrm{cc}([u < u(x) + \varepsilon], x) \nsubseteq H$, we get $\mathrm{cc}([u < u(x) + \varepsilon], x) \cap \mathrm{cc}([u <$

$\lambda]) \neq \emptyset$, which, in turn, implies that $\text{cc}([u < \lambda] \subseteq \text{cc}([u < u(x) + \varepsilon], x)$ and, thus, $\lambda \leq u(x) + \varepsilon$. The above results allow to conclude in the two possible configurations: $S_0 = \overline{\Omega}$ or $S_0 \subsetneq \overline{\Omega}$.

If $S_0 = \overline{\Omega}$, we have that $S^* \subsetneq \overline{\Omega}$, implying (see above) that $\lambda^* > u(x)$ and we easily derive

$$(\lambda^*, S_{\lambda^* - u(x)}) \preccurlyeq (\lambda^*, S^*).$$

Let $\varepsilon > 0$, and let $\varepsilon' = \min(\varepsilon, \lambda^* - u(x))$. Since $S_{\varepsilon'} \subsetneq \overline{\Omega}$, if $(\lambda, S) \preccurlyeq (u(x) + \varepsilon', S_{\varepsilon'})$, the above results prove that $\lambda \leq u(x) + \varepsilon'$, and, since $S \subsetneq \overline{\Omega}$, we have that $\lambda > u(x)$. Therefore $|\lambda - u(x)| \leq \varepsilon' \leq \varepsilon$, proving Equation (2.6).

Now, we assume that $S_0 \subsetneq \overline{\Omega}$. If, for some $\varepsilon > 0$, we have that $S_\varepsilon \subsetneq \overline{\Omega}$, we conclude in the same manner as in the previous case. Otherwise, let $(\lambda, S) \preccurlyeq (u(x), S_0)$. Observe that $\lambda \geq u(x)$. If $S = \text{Sat}(\text{cc}([u < \lambda]))$ is of lower type, necessarily $\lambda = u(x)$ and x is in a hole H of $\text{cc}([u < \lambda])$. Therefore $\text{cc}([u \geq u(x)], x) \subseteq H \subsetneq S$, which contradicts $(\lambda, S) \preccurlyeq (u(x), S_0)$. Then $S = \text{Sat}(C)$ is of upper type. If $\lambda > u(x)$, then x is in a hole H of C, and, thanks to Lemma 2.30, there is some lower shape S' at level λ contained in H and containing x. This easily implies that $S_{\lambda - u(x)} \subseteq S' \subsetneq \overline{\Omega}$, contradicting our previous assumption. In conclusion, $(\lambda, S) \preccurlyeq (u(x), S_0)$ implies that $\lambda = u(x)$ (and $S = S_0$), proving Equation (2.6), the infimum being actually reached at $(u(x), S_0)$. □

2.4.2 Indirect Reconstruction

The formulae for reconstruction we have established above are complex in the sense that they involve a limit. We provide here a simpler, algebraic reconstruction, in two steps: first, the level sets are deduced from the tree, then the reconstruction of the image is straightforward, given by

$$u(x) = \sup\{\lambda, x \in [u \geq \lambda]\} = \inf\{\lambda, x \in [u < \lambda]\}.$$

The reconstruction will derive from the following lemma.

Lemma 2.32. *Let $x \in \overline{\Omega}$ and $\lambda \in \mathbb{R}$. Then $\bigcap_{A \in \text{LS}_{\lambda,x}} A \in \text{LS}_{\lambda,x}$. More precisely, if $u(x) \geq \lambda$ then $\bigcap_{A \in \text{LS}_{\lambda,x}} A \in F_{\lambda,x}$, while if $u(x) < \lambda$, $\bigcap_{A \in \text{LS}_{\lambda,x}} A \in G_{\lambda,x}$.*

Proof. Suppose first that $\lambda > u(x)$ and define $C = \text{cc}([u < \lambda], x)$. We show that $\bigcap_{A \in \text{LS}_{\lambda,x}} A = \text{Sat}(C)$, which is sufficient as $\text{Sat}(C) \in G_{\lambda,x}$. Indeed, if $\text{Sat}(C') \in \text{LS}_{\lambda,x}$ then $\text{Sat}(C)$ and $\text{Sat}(C')$ are nested since they intersect at x. If $C \neq C'$, we have $C \cap C' = \emptyset$ since either C' is a distinct connected component of lower level set or $C' \subseteq [u \geq \lambda]$. Thus C is in an internal hole of C', so as $\text{Sat}(C)$ and we conclude $\text{Sat}(C) \subseteq \text{Sat}(C')$.

If $\lambda \leq u(x)$, we define $C = \mathrm{cc}([u \geq \lambda], x)$ and the same reasoning yields $\mathrm{Sat}(C) = \bigcap_{A \in \mathrm{LS}_{\lambda,x}} A$. □

Theorem 2.33. *We have the following representation formulae*

$$[u \geq \lambda] = \left\{ x \in \overline{\Omega},\ \bigcap_{A \in \mathrm{LS}_{\lambda,x}} A \in F_{\lambda,x} \right\} = \bigcup_{x \in \overline{\Omega}} \bigcup_{F \in F_{\lambda,x}} (F \setminus \bigcup_{G \in G_{\lambda,x}, G \subsetneq F} G)$$

and

$$[u < \lambda] = \left\{ x \in \overline{\Omega},\ \bigcap_{A \in \mathrm{LS}_{\lambda,x}} A \in G_{\lambda,x} \right\} = \bigcup_{x \in \overline{\Omega}} \bigcup_{G \in G_{\lambda,x}} (G \setminus \bigcup_{F \in F_{\lambda,x}, F \subsetneq G} F).$$

Proof. Let x be such that $u(x) \geq \lambda$, and $F = \mathrm{Sat}(\mathrm{cc}([u \geq \lambda], x))$. We have proved in Lemma 2.32 that $F = \bigcap_{A \in \mathrm{LS}_{\lambda,x}} A$. Conversely, let $x \in \overline{\Omega}$ be such that $F = \bigcap_{A \in \mathrm{LS}_{\lambda,x}} A \in F_{\lambda,x}$. We write $F = \mathrm{Sat}(C)$ with $C \in \mathcal{CC}([u \geq \lambda])$. If x were in an internal hole of C, thanks to Lemma 2.30 there would be some $G \in G_{\lambda,x}$ such that $G \subsetneq F$, contradicting the minimality of F. Thus $x \in C \subseteq [u \geq \lambda]$.

We have proved the first equality in the first representation formula. The other equalities are direct consequences of this one. □

In the discrete setting, where images are constant on each pixel (Chapter 6 will provide detailed definition), the formulae of Theorem 2.33 provide a constructive recovery of the level sets of u from the shapes since there are a finite number of distinct shapes. The right hand side of these formulae represent partitions of $\mathcal{CC}([u \geq \lambda])$ and $\mathcal{CC}([u < \lambda])$ into their connected components. For example the set $F \setminus \bigcup_{G \in G_{\lambda,x}, G \subsetneq F} G$ is $F = \mathrm{Sat}(C)$ with $C \in \mathcal{CC}([u \geq \lambda])$ minus the holes of C.

2.5 Finiteness of the Tree for Images with Grains of Minimal Positive Size

Definition 2.34. We say that an image u has grains of minimal positive size if there exists some $\delta > 0$ such that for any connected component K of an upper or lower level set of u, $|K| \geq \delta$.

After application of an area opening and an area closing filters of area δ (see [114, 115]), any image verifies this property (see [11]). Provided such an image, we will show that the tree, whereas it can have an infinite number of nodes, actually has a finite structure. **In this whole section, we will suppose that u is an image with grains of minimal positive size δ.**

Lemma 2.35. *For each $\lambda \in \mathbb{R}$ there is a finite number of connected components of $[u \geq \lambda]$ and each component has a finite number of holes.*

Proof. Let $\lambda \in \mathbb{R}$. Since each connected component of $[u \geq \lambda]$ has area $\geq \delta$, there must be a finite number of them. Let Y be a component of $[u \geq \lambda]$ and let H be a hole of Y. Since H is open, H cannot be covered by components of $[u \geq \lambda]$. Hence $H \cap [u < \lambda] \neq \emptyset$. We conclude that each hole of Y contains a component of $[u < \lambda]$. Hence there may be only a finite number of them.□

Lemma 2.36. *(i) Let T be any shape. Then T contains a connected compo nent of an upper or lower level set, hence $|T| \geq \delta$.*
(ii) Let $S \subseteq T$ be two shapes of different type. Then T contains a connected component of an upper or lower level set, hence $|T \setminus S| \geq \delta$.

Proof. (i) Since any shape T contains a connected component of an upper or lower level set, then $|T| \geq \delta$.

(ii) Suppose that $S = \mathrm{Sat}(X)$, $T = \mathrm{Sat}(Y)$, where X and Y are connected components of level sets of u. If $X \cap Y = \emptyset$, then $Y \subseteq T \setminus S$, thus, $|T \setminus S| \geq |Y| \geq \delta$. Therefore, we may assume that $X \cap Y \neq \emptyset$.

Suppose first that $X = \mathrm{cc}([u \geq \lambda])$ and $Y = \mathrm{cc}([u < \mu])$, $\lambda, \mu \in \mathbb{R}$. As $X \cap Y \neq \emptyset$, $\lambda < \mu$. Assume that $Y \setminus S \subseteq [u \geq \lambda]$. In this case, we could write

$$Y = (Y \cap S) \cup (Y \cap ([u \geq \lambda] \setminus S)) .$$

Now, S being closed, $Y \cap S$ is a closed set of Y, and $[u \geq \lambda] \setminus S$, being a finite union of connected components of $[u \geq \lambda]$, is also closed. Hence, we have a partition of the set Y into two closed subsets, contradicting its connectedness. Therefore, $Y \cap ([u < \lambda] \setminus S) \neq \emptyset$. In particular, $Y \setminus S$ contains a connected component of $[u < \lambda]$, hence, we have that

$$|T \setminus S| \geq |Y \setminus S| \geq \delta. \tag{2.8}$$

In the same way, if $X = \mathrm{cc}([u < \lambda])$ and $Y = \mathrm{cc}([u \geq \mu])$, we have that $\mu < \lambda$, and, since Y is connected and meets both S and $\overline{\Omega} \setminus S$, we also have that $Y \cap \partial S \neq \emptyset$. Since $\partial S \subseteq \partial X \subseteq [u \geq \lambda]$ and $S \cap \partial S = \emptyset$, we have that $(Y \setminus S) \cap [u \geq \lambda] \neq \emptyset$, and, therefore, $Y \setminus S$ contains a connected component of $[u \geq \lambda]$, yielding (2.8). □

Proposition 2.37. *There is a finite number of leaves in the tree of shapes of u.*

Proof. Since any leaf T is an intersection of a nested family of shapes, which, by Lemma 2.36 (i), are all of measure $\geq \delta$, then $|T| \geq \delta$. Since, by Proposition 2.26, any two distinct leaves are disjoint, we conclude that there must be a finite number of them. □

Theorem 2.38. *There is a finite number of maximal monotone sections in the tree of shapes of u.*

Proof. Suppose that there is an infinite number of maximal monotone sections $[S_n^*, T_n^*]$ in the tree of shapes of u. Without loss of generality, we may assume that all are of the same type. Then $[S_i^*, T_i^*] \cap [S_j^*, T_j^*] = \emptyset$, for all $i \neq j$. Let us consider S_i shapes such that $S_i \subseteq [S_i^*, T_i^*]$ for all $i \geq 1$. If there is an infinite set of indices $I \subseteq \mathbb{N}$ such that $S_i \cap S_j = \emptyset$ for all $i, j \in I$, then by Lemma 2.36 (i), $\overline{\Omega}$ would be of infinite measure. Thus there is only a finite number of indices $i \in \mathbb{N}$ such that the S_i are two by two disjoint. Thus there is i_1 such that S_{i_1} contains an infinity of S_i. Without loss of generality, we may assume that $i_1 = 1$ and all $S_i \subseteq S_1$ for $i \geq 2$. By the same reasons there are at most a finite number of indices $i \geq 2$ such that S_i are two by two disjoint. Thus, there is an index i_2, which we may assume to be $i_2 = 2$, such that $S_2 \subseteq S_1$ and S_2 contains an infinity of S_i. In this way, we construct a family of sets S_n such that $S_{n+1} \subseteq S_n$ for all $n \geq 1$.

Let us consider the interval $[S_{n+1}, S_n]$. We claim that $|S_n \setminus S_{n+1}| \geq \delta$. If there is a shape S such that $S \subseteq S_n$ and $S \cap S_{n+1} = \emptyset$, then $S \subseteq S_n \setminus S_{n+1}$, and, by Lemma 2.36 (i), $|S_n \setminus S_{n+1}| \geq \delta$. Thus, we may assume that there is no bifurcation in $[S_{n+1}, S_n]$. Suppose that all shapes in $[S_{n+1}, S_n]$ are of the same type. Thus, $[S_{n+1}, S_n]$ is a monotone section with all shapes of the same type. In this case, by Proposition 2.28, $[S_{n+1}, S_n]$ would be a monotone section intersecting both $[S_{n+1}^*, T_{n+1}^*]$ and $[S_n^*, T_n^*]$, and there would exist a maximal monotone section containing both of them. This contradiction proves that there is a shape $Q_n \in [S_{n+1}, S_n]$ of different type as the ones of the maximal monotone sections that we are considering. By Lemma 2.36 (ii), we have that $|S_n \setminus Q_n| \geq \delta$. Thus, also $|S_n \setminus S_{n+1}| \geq \delta$. Thus, $|S_n \setminus S_{n+1}| \geq \delta$ for all $n \geq 1$. Now, since $(S_{n-1} \setminus S_n) \cap (S_n \setminus S_{n+1}) = \emptyset$ for all $n \geq 1$, this implies that $\overline{\Omega}$ must have infinite measure. This contradiction implies the statement of the Theorem. □

Corollary 2.39. *There is a finite number of maximal branches in the tree of shapes of u.*

Proof. Each maximal branch containing a maximal monotone section, their number cannot exceed the number of the maximal monotone sections, thus, it is finite. □

Remark 2.40. Proposition 2.37, Theorem 2.38, and Corollary 2.39 also hold for continuous functions with a finite number of extrema. This will become clear in Chap. 6 where all precise definitions are given.

2.6 Applications

2.6.1 A Simple Application: Adaptive Quantization

The tree of shapes can be used to give quantized versions of an image. This can be done in two steps: first, remove small shapes of the tree, then keep in each remaining maximal monotone section only the most significant shape.

The filtered image is obtained by reconstruction of the resulting tree. The first step corresponds to the grain filter, as studied in the BV framework in [6], and is a generalization of the area opening and area closing, as introduced in [114,115]. The grain filter removes shapes of area smaller than a parameter a, whereas area opening (resp. closing) removes connected components of upper (resp. lower) level sets of area $\leq a$. The grain filter has the advantage over these filters to act simultaneously on upper and lower level sets. Moreover, it is self-dual on continuous functions as it was proved in [11] or in [67] for smooth functions of class C^N, N being the dimension of the space. Thus this filter permits to simplify the topographic map of the image while keeping the geometric information of the image, given in terms of its 'level lines' or its shapes, and do it in a symmetric way with respect to the upper and lower level sets, at least when the image is a continuous function. This prefiltering step is crucial in many image processing tasks, like morphological compression or registration, to mention two of them. The tree of shapes is well adapted to compute the action of the grain filter on an image in a simple and efficient way. Detailed analysis of grain filters is done in Chap. 3.

The second step works on maximal monotone sections: in each such section, we remove all shapes except the most significant one. The meaningfulness of a shape is chosen as the ratio A/L^2, where L is the length of the boundary and A the area. Thus the most round shapes remain. However, any other criterion that seems appropriate can be used, as for example the minimum gradient norm along the boundary. The remaining shape gets the mean gray level of the section, so as not to change the local contrast.

The idea behind this is twofold: small shapes represent only details of the image, or are mainly due to noise; the maximal monotone sections represent visual "objects" present multiple times in the image due to the gradation at their border, which comes from the smoothing during the acquisition process. Moreover, the removal of small shapes can eliminate some bifurcations, yielding longer maximal monotone sections.

However, we must be careful because some shapes cannot be removed without affecting the reconstruction. It can happen that the reconstructed image's tree of shapes is not the one on which the reconstruction was based. This is illustrated in Fig. 2.4. To avoid this unfortunate effect, it is sufficient to prevent the removal of the minimal shape of each maximal monotone section if one of its children is of a different type.

Figure 2.5 shows some results obtained by this adaptive quantization.

2.6.2 Other Applications

We review in this section some applications of the tree of shapes in digital image processing. In these applications the authors use the fast algorithm

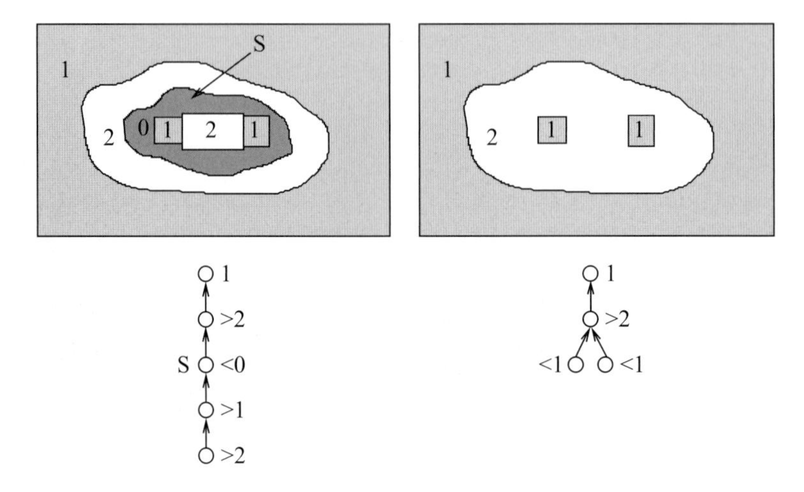

Fig. 2.4 The removal of the shape S should be prevented: the reconstructed image's tree structure is not the modified tree.

extracting the tree of shapes as presented in [79]. Several of these applications will be discussed in more detail in Chap. 8.

Desolneux *et al.* [25] use the tree of shapes to extract the level lines of the image with high contrast, in other words, the edges. Level lines having a minimal contrast above some threshold, which is based on statistical considerations, are considered meaningful. However, frequently, it happens that when a meaningful level line is detected, also are many other level lines inside the same monotone section. To get rid of this redundancy, they keep only the most meaningful level lines in a monotone section. This simplification of the tree is very close to the adaptive quantization presented above, with the addition of a parameter measuring the meaningfulness.

Gousseau [37] uses the shapes to reproduce textures. Many shapes inside a texture are extracted and put in random locations in a blank image to synthesize the texture.

Dibos and Koepfler [27,29] use the tree of shapes to construct a denoising filter which does not coincide with the grain filter [11]. Indeed, to decrease the total variation of an image by alteration of its level lines there are essentially two possibilities: either make them shorter, usually by evolution of a PDE, or to decrease their contrast. The latter solution is adopted in [27].

Shapes can also be used for registration purposes [28,77]: some invariant characteristics of shapes are extracted in different images and compared to get correspondences. Voting procedures can then be used to recover global displacement parameters. A more complex but more accurate method to extract the correspondences is proposed in [58], where pieces of level lines are compared.

Fig. 2.5 Effect of the adaptive quantization on an image of size 500×500. From left to right, top to bottom: $a = 1$ (8,834 sections), $a = 20$ (1,938), $a = 400$ (155) and $a = 8{,}000$ (18).

2.7 Comparison with Component Tree

Let us compare the tree of shapes with the component trees, that is the tree of connected components of upper, or lower, level sets. Referring to Fig. 2.6, several semantic interpretations of the scene are possible. For example, B, C and D could be part of a single long band partially occluded by a disk A with a square hole; or the four of them may be independent objects. Because of this ambiguity, an ideal representation would be orthogonal to the interpretation. The output of a segmentation algorithm coupled with the region adjacency graph would correspond well to this requirement: the regions A, B, C and D would be separated but their adjacency coded. Unfortunately, in a complex scene, adjacent regions of an iso-level have to be merged, which

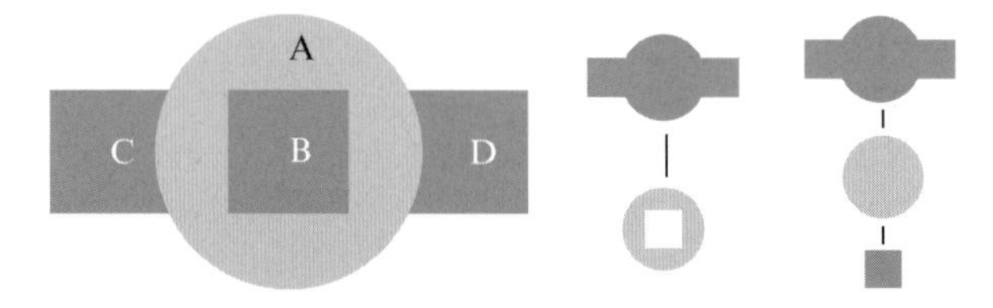

Fig. 2.6 Semantic interpretation provided of a scene given by trees. Left: image of a scene; center: max-tree; right: tree of shapes. The real root, corresponding to the full image, is omitted from both trees.

supposes a merging criterion and a merging order. It has to be recognized that despite a rich literature on image segmentation, no algorithm is really satisfactory in this regard.

Component trees and tree of shapes do this merging based on gray level comparison. In Fig. 2.6, the max-tree and the tree of shapes have the same root. Then the max tree extracts A as a disk with a square hole. By comparison, the tree of shapes extract $\mathrm{Sat}(A) = A \cup B$ as a node and B as another node. We would argue that the latter is more useful than the former representation. While B could always be recovered from A through hole extraction, this is explicit in the tree of shapes. Whereas C and D are always merged with A in both trees (which is a drawback since C and D not being adjacent, their gray levels should not be compared for merging if we want to be local contrast invariant), B is explicit in the tree of shapes, independently of C and D. For this reason, while both trees lead a semantic interpretation of the scene, the tree of shapes preserves better the original ambiguity of the scene.

Chapter 3
Grain Filters

Section 2.5 showed that if we ignore fine scales of the image, the tree of shapes is essentially finite. Natural candidates to eliminate these fine details are the grain filters. In the same manner as extrema filters act on upper and lower component trees, we present a similar filter acting on the tree of shapes and study its properties. This chapter is completely in the realm of mathematical morphology but with a somewhat unusual framework, since the domain of the image is not a discrete set of pixels but the continuous domain. In that case, regularity assumptions on the image are required.

3.1 Introduction

Filters used to simplify an image and satisfying a minimal set of invariance properties are scarce. Actually, only one of them has the maximal set of invariance properties, and it is driven by the parabolic partial differential equation [5, 76, 99]:

$$\frac{\partial u}{\partial t} = \|\nabla u\| \mathrm{curv}^{1/3} u,$$

where curv $u(x)$ is the curvature of the level line of u at the point x, this being restricted to the regular points of u.

However, the previous filter is optimal (in terms of invariance) among *regular* filters, that is, filters driven by a P.D.E. This property of regularity, while desirable in theory, has the drawback of modifying all contours, and, in particular, of destroying T-junctions, which are important clues for occlusion [17]. If we drop this requirement, a bunch of other filters satisfying the same invariance properties are available.

Motivated by the study of a family of filters by reconstruction [53, 54, 92, 114, 115], Serra and Salembier [98, 106] introduced the notion of connected operators. Such operators simplify the topographic map of the image. These filters have become very popular in image processing because, on an experimental basis, they have been claimed to simplify the image while preserving contours. This property has made them very attractive for a large number

V. Caselles and P. Monasse, *Geometric Description of Images as Topographic Maps*, Lecture Notes in Mathematics 1984,
DOI 10.1007/978-3-642-04611-7_3,

of applications, such as noise cancelation or segmentation [72, 116]. More recently, they have become the basis of a morphological approach to image and video compression [35,94,96,97]. Different classes of connected operators have been studied by Meyer [70, 71], Serra [105] or Heijmans [45] (see also references therein).

This chapter is devoted to study the theoretical properties of two kinds of connected operators: the extrema filters and the "shape" filters. Each of them simplifies the topographic map of the image, but with different senses given to the term topographic map. The maxima filter removes connected components of upper level sets of insufficient area, while keeping the other ones identical [114, 115]. This ensures that regional maxima of the filtered image have a minimal grain size. Similarly, the minima filter removes too small connected components of lower level sets. For these filters, the "grain" corresponds to a connected component of a level set, and small grains are considered as noise. This can be seen as the pruning of the tree of the connected components of upper, or lower, level sets.

In Chap. 2 we introduced the notion of "shapes", designed to deal symmetrically with upper and lower level sets. The shapes are also organized in a tree, driven by inclusion. When applied to images of positive minimal grain size, we showed that the structure of this tree is finite. As shown here, any image resulting from the application of the extrema filters has this property. This new tree also provides the definition of another grain filter, for which the grain is a shape [67, 79, 80]. It removes small shapes while preserving the ones of sufficient area. The essential improvement over the extrema filters is that it deals in the same manner with upper and lower level sets. In the vocabulary of mathematical morphology [69, 102, 103], this filter is self-dual when applied to continuous functions.

The present chapter is organized as follows. Section 3.2 recalls the foundations of mathematical morphology and underlines the link between contrast invariant operators on images and set operators. Section 3.3 introduces the main properties of extrema filters and proves them. In Sect. 3.4 we prove the analogous properties for the grain filter and, in particular, that it is a self-dual filter on continuous functions, generalizing a result of [6]. Finally, in Sect. 3.6 we illustrate these filters with an experiment.

3.2 General Results From Mathematical Morphology

This section is a review of some well known results of Mathematical Morphology. We review them in some detail for the reader's convenience. More detailed accounts on them can be seen in [42, 43, 64, 65], or [39]. Throughout this section, we shall consider real functions defined in a subset of $\mathbb{R}^N$. As in Chap. 2, if u is a real function defined on $D \subseteq \mathbb{R}^N$, we denote by $[u \geq \lambda]$ the set $\{x \in D : u(x) \geq \lambda\}$, $\lambda \in \mathbb{R}$. Similarly, we define the sets $[u > \lambda]$, $[u \leq \lambda]$, $[u < \lambda]$.

3.2.1 Level Sets and Contrast Invariance

Let $D \subseteq \mathbb{R}^N$. If $u : D \to \mathbb{R}$ is any real function, and $\mathcal{X}_\lambda u = [u \geq \lambda]$, then

$$\forall \lambda \in \mathbb{R}, \quad \mathcal{X}_\lambda u = \bigcap_{\mu < \lambda} \mathcal{X}_\mu u$$

which implies, in particular, that

$$\forall \lambda \geq \mu, \quad \mathcal{X}_\lambda u \subseteq \mathcal{X}_\mu u.$$

Conversely, if X_λ is a family of sets satisfying

$$\forall \lambda, \quad X_\lambda = \bigcap_{\mu < \lambda} X_\mu$$

then X_λ is the level set at level λ of the function u defined by

$$u(x) = \sup\{\lambda, x \in X_\lambda\}$$

namely, $\mathcal{X}_\lambda u = X_\lambda$.

Under the weaker hypothesis of monotonicity of $\{X_\lambda\}_{\lambda \in \mathbb{R}}$, Guichard and Morel show in [39] that $\mathcal{X}_\lambda u = X_\lambda$ a.e. and for almost every $\lambda \in \mathbb{R}$.

A contrast change, in the restrictive sense, is a strictly increasing continuous map $g : \mathbb{R} \to \mathbb{R}$. It is therefore a homeomorphism of $\mathbb{R}$ onto an open interval of $\mathbb{R}$. A direct consequence is

$$\forall \lambda, \mu \in \mathbb{R}, \quad g(\lambda) \geq \mu \Leftrightarrow \lambda \geq g^{-1}(\mu). \tag{3.1}$$

Such relation is called in mathematical morphology an adjunction [38, 43].

For an image $u : D \to \mathbb{R}$, the contrast change g applied to u is $g \circ u$. A contrast change g will indifferently be considered as a function defined on $\mathbb{R}$ or as an operator acting on functions u. A direct consequence of (3.1) is

$$[g \circ u \geq \lambda] = [u \geq g^{-1}(\lambda)],$$

which can also be written as

$$\mathcal{X}_\lambda\, g \circ u = \mathcal{X}_{g^{-1}(\lambda)} u,$$

showing that the families of level sets of $g \circ u$ and of u are the same, only their level changes.

To deal with operators acting on upper semicontinuous functions it will be useful to have a more general class of contrast changes. Recall that a function $u : X \to \mathbb{R}$ defined on any topological space X is called upper semicontinuous if, for any $\lambda \in \mathbb{R}$, the upper level set $[u \geq \lambda]$ is closed. We say that $g : \mathbb{R} \to \mathbb{R}$ is a *general contrast change* if g is nondecreasing and upper semicontinuous. For a general contrast change, we have that

$$g(\lambda) \geq \mu \quad \Leftrightarrow \quad \lambda \geq g^{(-1)}(\mu)$$

where

$$g^{(-1)}(x) = \inf\{y : \ g(y) \geq x\} = \min\{y : \ g(y) \geq x\}.$$

We use the convention $\inf \emptyset = -\infty$.

3.2.2 Link Between Set Operator and Contrast Invariant Operators on Images

Let $D \subseteq \mathbb{R}^N$. If T is an increasing operator acting on subsets of D (i.e., if $X \subseteq Y$, then $T(X) \subseteq T(Y)$), a necessary requirement for T to transform the level sets of a function into the level sets of a function is thus

$$\forall F_0 \supseteq F_1 \supseteq \cdots \supseteq F_n \supseteq \dots, \quad T(\bigcap_n F_n) = \bigcap_n T(F_n). \tag{3.2}$$

The filter $\tilde{T}$ associated to T, and acting on functions $u : D \to \mathbb{R}$ is defined by

$$\forall \lambda \in \mathbb{R}, \quad [\tilde{T}u \geq \lambda] = T([u \geq \lambda]),$$

or equivalently

$$\tilde{T}u(x) = \sup\{\lambda, x \in T([u \geq \lambda])\}.$$

If we denote by $B_x = \{B : \ x \in T(B)\}$, then we observe ([64, 65]) that

$$\tilde{T}u(x) = \sup_{B \in B_x} \inf_{y \in B} u(y).$$

Indeed, let $\tilde{T}'u(x)$ be the right hand side term of this equality. If λ is such that $x \in T([u \geq \lambda])$, by definition of B_x we deduce that $[u \geq \lambda] \in B_x$; on the other hand, since

$$\inf_{y \in [u \geq \lambda]} u(y) \geq \lambda$$

we get immediately $\tilde{T}'u(x) \geq \tilde{T}u(x)$. Conversely, if $\lambda = \tilde{T}'u(x)$ then

$$\forall n > 0, \exists B_n \in B_x, \quad \inf_{y \in B_n} u(y) \geq \lambda - \frac{1}{n}.$$

That is, $B_n \subseteq [u \geq \lambda - 1/n]$ and, therefore,

$$x \in \tilde{T}([u \geq \lambda - \frac{1}{n}]).$$

Taking the intersection over all n, we get that $x \in \tilde{T}([u \geq \lambda])$, i.e., $\tilde{T}u(x) \geq \lambda = \tilde{T}'u(x)$.

Since the set operator T is increasing, the operator $\tilde{T}$ on images is also increasing (or order preserving), i.e., $\tilde{T}u \leq \tilde{T}v$ when u, v are images such that $u \leq v$. The word filter in mathematical morphology is reserved for increasing and idempotent operators.

If T is only defined on closed subsets of D and satisfies (3.2) when the sets F_n are closed, then $\tilde{T}$ can be defined on upper semicontinuous functions $u : D \to \mathbb{R}$. Then, if $g : \mathbb{R} \to \mathbb{R}$ is any general contrast change, it is easy to check that $\tilde{T}$ and g commute when applied to upper semicontinuous functions. We get contrast invariance in a strong sense, since g can have constant stretches, and needs not be continuous.

Let us denote by $\mathcal{USC}(D)$ (resp., $C(D)$) the family of upper semicontinuous (resp., continuous) functions on D.

Definition 3.1. A contrast invariant operator $\tilde{T}$ on $\mathcal{USC}(D)$ is a map acting on functions $u \in \mathcal{USC}(D)$ that commutes with any general contrast change: $g \circ \tilde{T} = \tilde{T} \circ g$.

Remark. Similarly, we may define contrast invariant operators acting on any family of functions $\mathcal{F}$, the only requirement being that $g \circ u \in \mathcal{F}$ when $u \in \mathcal{F}$ and g is any general contrast change. Since, for the purposes of this paper we only need upper semicontinuous functions we shall restrict our definition to them.

3.3 Extrema Filters

3.3.1 Definition

Extrema filters are constructed in such manner that the connected components of level sets of an image have a minimum area. We call them extrema filters because a connected component of level set contains a regional extremum. This is achieved in two steps: first, connected components of upper level sets are filtered, then lower level sets. We define the set operators ensuring such properties.

Let us first fix some notation. Let $\overline{\Omega}$ be a set homeomorphic to the closed unit ball of $\mathbb{R}^N$ $(N \geq 2)$, $\{x \in \mathbb{R}^N, \|x\| \leq 1\}$, and Ω be the interior of $\overline{\Omega}$. Note that, in particular, $\overline{\Omega}$ is compact, connected and locally connected. Moreover, $\overline{\Omega}$ is unicoherent. We shall use the notation introduced in Sect. 2.2 of Chap. 2.

Given a measurable subset X of $\overline{\Omega}$, we denote by $|X|$ the Lebesgue measure of X. Observe that if X is measurable, the connected components of X, being closed sets in X, are also measurable.

Let ε be a parameter, representing an area threshold, and let X be a measurable subset of $\overline{\Omega}$. We define the filters

$$M_\varepsilon(X) = \bigcup \{C \in \mathcal{CC}(X) : |C| \geq \varepsilon\},$$
$$M'_\varepsilon(X) = \bigcup \{C \in \mathcal{CC}(X) : |C| > \varepsilon\}.$$

We shall observe below that both operators M_ε and M'_ε, which are well known in mathematical morphology [114, 115], are increasing and idempotent, thus they are filters. Since they are also anti-extensive they are openings, and they are called area openings. They may be used to define area closings by computing the corresponding negative operator, for instance, the negative of M'_ε would be defined by $M_\varepsilon^{'*}(X) = \overline{\Omega} \setminus M'_\varepsilon(\overline{\Omega} \setminus X)$. Then $M_\varepsilon^{'*}$ is increasing, idempotent and extensive, thus, it is a closing, called the area closing [38, 114, 115]. Let $B_\varepsilon = \{B : B$ measurable and connected, $0 \in B$, $|B| \geq \varepsilon\}$ and $B'_\varepsilon = \{B : B$ measurable and connected, $0 \in B$, $|B| > \varepsilon\}$. We define $x + B_\varepsilon = \{x + B : B \in B_\varepsilon\}$, $x + B'_\varepsilon = \{x + B : B \in B'_\varepsilon\}$ where $x + B = \{x + y : y \in B\}$. We define the maxima filter M_ε^+ and the minima filter M_ε^- on a function $u : \overline{\Omega} \to \mathbb{R}$ by

$$M_\varepsilon^+ u = \sup_{B \in x + B_\varepsilon} \inf_{y \in B} u(y),$$
$$M_\varepsilon^- u = \inf_{B \in x + B'_\varepsilon} \sup_{y \in B} u(y).$$

Again, these operators are well known in mathematical morphology. As we shall observe below, both are increasing and idempotent, thus they are filters. $M_\varepsilon^+ u$ comes from applying M_ε to upper level sets u, thus, it is an anti-extensive filter, in other words, an opening. $M_\varepsilon^- u$ comes from applying M'_ε to the lower level sets of u or, equivalently, from applying $M_\varepsilon^{'*}$ to the upper level sets of u, thus is an extensive filter, or, in other words, a closing. The definitions are voluntarily not symmetric, so that both can act on (upper) semicontinuous functions. We will show that, when defined on upper semicontinuous functions, M_ε^+ and M_ε^- are contrast invariant operators whose associated set operators are M_ε and M'_ε, respectively. To avoid the cases where $B_\varepsilon = \emptyset$ or $B'_\varepsilon = \emptyset$, we will always suppose that $\varepsilon < |\overline{\Omega}|$.

3.3.2 Preliminary Results

Lemma 3.2. *If $(C_n)_{n \in \mathbb{N}}$ is a nonincreasing sequence of compact sets and $C = \bigcap_{n \in \mathbb{N}} C_n$, then*

$$\mathrm{cc}(C,x) = \bigcap_{n\in\mathbb{N}} \mathrm{cc}(C_n,x).$$

If $(O_n)_{n\in\mathbb{N}}$ is a nondecreasing sequence of open sets and $O = \bigcup_{n\in\mathbb{N}} O_n$, then

$$\mathrm{cc}(O,x) = \bigcup_{n\in\mathbb{N}} \mathrm{cc}(O_n,x).$$

Proof. It is clear that $\mathrm{cc}(C,x) \subseteq \mathrm{cc}(C_n,x)$ for any n. Conversely, $\bigcap_{n\in\mathbb{N}} \mathrm{cc}(C_n,x)$ is an intersection of continua, thus, it is a continuum. Since it contains x and is included in C, we get the other inclusion.

For any n, $\mathrm{cc}(O_n,x) \subseteq O$, thus $\bigcup_{n\in\mathbb{N}} \mathrm{cc}(O_n,x) \subseteq \mathrm{cc}(O,x)$. On the other hand, O being open and $\overline{\Omega}$ locally connected, $\mathrm{cc}(O,x)$ is an open set. Hence, for any $y \in \mathrm{cc}(O,x)$, there is some continuum $K_y \subseteq \mathrm{cc}(O,x)$ containing x and y. Since $K_y \subseteq \bigcup_{n\in\mathbb{N}} O_n$ and it is a compact set, we can extract a finite covering of K_y, and as the sequence (O_n) is nondecreasing, there is some n such that $K_y \subseteq O_n$. Since K_y is connected and contains x, we have that $y \in K_y \subseteq \mathrm{cc}(O_n,x)$. We conclude that $\mathrm{cc}(O,x) \subseteq \bigcup_{n\in\mathbb{N}} \mathrm{cc}(O_n,x)$. □

Proposition 3.3. *We have the following properties for M_ε and M'_ε:*

(i) M_ε and M'_ε are nondecreasing on measurable subsets of $\overline{\Omega}$.
(ii) M_ε is upper semicontinuous on compact sets: $(F_n)_{n\geq 0}$ being a nonincreasing sequence of compact sets, then $M_\varepsilon(\bigcap_n F_n) = \bigcap_n M_\varepsilon(F_n)$.
(iii) M'_ε is lower semicontinuous on open sets: $(O_n)_{n\geq 0}$ being a nondecreasing sequence of open sets, then $M'_\varepsilon(\bigcup_n O_n) = \bigcup_n M'_\varepsilon(O_n)$.

Proof. Property (i) is a direct consequence of the definitions.
(ii) Let $F = \bigcap_{n\in\mathbb{N}} F_n$. Since M_ε is monotone, we have that $M_\varepsilon(F) \subseteq \bigcap_{n\in\mathbb{N}} M_\varepsilon(F_n)$. Now, let $x \in M_\varepsilon(F_n)$ for all n, and $C_n = \mathrm{cc}(F_n,x)$. Then $|C_n| \geq \varepsilon$. Let $C = \bigcap_{n\in\mathbb{N}} C_n$. Applying Lemma 3.2, we observe that $C = \mathrm{cc}(F,x)$. In particular, $|C| = \inf |C_n| \geq \varepsilon$ and we deduce $x \in C \subseteq M_\varepsilon(F)$.
(iii) Let $O = \bigcup_n O_n$. Since M'_ε is monotone, $\bigcup_n M'_\varepsilon(O_n) \subseteq M'_\varepsilon(O)$. Now, let $x \in M'_\varepsilon(O)$. Then $U = \mathrm{cc}(O,x)$ is such that $|U| > \varepsilon$. Let $U_n = \mathrm{cc}(O_n,x)$. Lemma 3.2 proves that

$$U = \bigcup_{n\in\mathbb{N}} U_n.$$

Then $\sup |U_n| = |U| > \varepsilon$. Hence for n large enough, $|U_n| > \varepsilon$. We conclude that $x \in U_n \subseteq M'_\varepsilon(O_n)$. □

If $\mathcal{A}$ and $\mathcal{B}$ are two families of sets, we say that $\mathcal{A}$ is a basis of $\mathcal{B}$ if $\mathcal{A} \subseteq \mathcal{B}$ and for any $B \in \mathcal{B}$, there is some $A \in \mathcal{A}$ such that $A \subseteq B$.

Lemma 3.4. *B_ε is a basis of $\{X : 0 \in M_\varepsilon(X)\}$ and B'_ε is a basis of $\{X : 0 \in M'_\varepsilon(X)\}$.*

Proof. This is a direct consequence of the definitions. □

Corollary 3.5. *Applied to upper semicontinuous functions, M_ε^+ (resp. M_ε^-) is a contrast invariant operator whose associated set operator is M_ε (resp. M'_ε). More precisely, for all λ,*

$$[M_\varepsilon^+ u \geq \lambda] = M_\varepsilon([u \geq \lambda]);$$
$$[M_\varepsilon^- u < \lambda] = M'_\varepsilon([u < \lambda]).$$

Proof. Let $C_\varepsilon = \{X : 0 \in M_\varepsilon(X)\}$. As a consequence of Proposition 3.3, we have

$$\left\{x : \sup_{B \in x + C_\varepsilon} \inf_{y \in B} u(y) \geq \lambda\right\} = M_\varepsilon([u \geq \lambda]).$$

We now use the fact that B_ε is a basis of C_ε, as shown in Lemma 3.4. As $B_\varepsilon \subseteq C_\varepsilon$, we deduce that

$$M_\varepsilon^+ u\,(x) \leq \sup_{B \in x + C_\varepsilon} \inf_{y \in B} u(y).$$

If $B \in x + C_\varepsilon$, there is some $B' \in x + B_\varepsilon$ such that $B' \subseteq B$. Then

$$\inf_{y \in B} u(y) \leq \inf_{y \in B'} u(y) \leq M_\varepsilon^+ u\,(x),$$

and by taking the supremum over all B, we get

$$\sup_{B \in x + C_\varepsilon} \inf_{y \in B} u(y) \leq M_\varepsilon^+ u\,(x).$$

A similar proof applies to link M_ε^- and M'_ε. □

When we write the inequality $u \leq v$ for two functions $u, v : \overline{\Omega} \to \mathbb{R}$ we mean that $u(x) \leq v(x)$ for all $x \in \overline{\Omega}$. We shall write $\|u\|_\infty := \sup_{x \in \overline{\Omega}} |u(x)|$.

Proposition 3.6. *Let $u, v : \overline{\Omega} \to \mathbb{R}$. Then*

(i) If $\varepsilon < \varepsilon'$ then $M_{\varepsilon'}^+ u \leq M_\varepsilon^+ u \leq u \leq M_\varepsilon^- u \leq M_{\varepsilon'}^- u$.
(ii) If $u \leq v$, then $M_\varepsilon^+ u \leq M_\varepsilon^+ v$ and $M_\varepsilon^- u \leq M_\varepsilon^- v$.
(iii) $M_\varepsilon^\pm(\alpha) = \alpha$ for all $\alpha \subset \mathbb{R}$.
(iv) $M_\varepsilon^\pm(u + \alpha) = M_\varepsilon^\pm u + \alpha$ for all $\alpha \in \mathbb{R}$.
(v) $\|M_\varepsilon^\pm(u) - M_\varepsilon^\pm(v)\|_\infty \leq \|u - v\|_\infty$

The proofs of $(i) - (iv)$ are immediate and we will not include the details. Let us briefly give the proof of (v). If $\|u - v\|_\infty = \infty$, we are done. Assume that $\alpha := \|u - v\|_\infty < \infty$. In this case $u \leq v + \alpha$. Then from (ii) and (iv) we deduce that $M_\varepsilon^\pm(u) \leq M_\varepsilon^\pm(v) + \alpha$. Interchanging the roles of u and v we obtain the other inequality required to complete the proof of (v).

Corollary 3.7. *Let $u_n, u : \overline{\Omega} \to \mathbb{R}$ be such that $u_n \to u$ uniformly in $\overline{\Omega}$. Then $M_\varepsilon^\pm u_n \to u$ as $n \to \infty$ uniformly in $\overline{\Omega}$.*

3.3.3 Properties

Proposition 3.8. *If u is an upper semicontinuous function, so are $M_\varepsilon^+ u$ and $M_\varepsilon^- u$. If u is continuous, so are $M_\varepsilon^+ u$ and $M_\varepsilon^- u$.*

Proof. Let u be an upper semicontinuous function. Then, for any λ, $[M_\varepsilon^+ u \geq \lambda] = M_\varepsilon([u \geq \lambda])$. As $[u \geq \lambda]$ is a closed set, its connected components are closed. Since $M_\varepsilon([u \geq \lambda])$ is a finite union of some of them, it is closed. Thus $M_\varepsilon^+ u$ is upper semicontinuous.

In the same manner, $[M_\varepsilon^- u < \lambda] = M_\varepsilon'([u < \lambda])$. Since $[u < \lambda]$ is an open set, its connected components are also open, and its image by M_ε' is thus a union of open sets, which is open. This proves that $M_\varepsilon^- u$ is upper semicontinuous.

Finally, suppose that u is continuous. We just have to prove that $[M_\varepsilon^\pm u \leq \lambda]$ is closed for any $\lambda \in \mathbb{R}$. Using Corollary 3.5, we write

$$\begin{aligned}
[M_\varepsilon^+ u \leq \lambda] &= \bigcap_{n>0} [M_\varepsilon^+ u < \lambda + \frac{1}{n}] \\
&= \bigcap_{n>0} (\overline{\Omega} \setminus [M_\varepsilon^+ u \geq \lambda + \frac{1}{n}]) \\
&= \bigcap_{n>0} (\overline{\Omega} \setminus M_\varepsilon([u \geq \lambda + \frac{1}{n}])).
\end{aligned}$$

We claim that

$$\bigcup_{n>0} M_\varepsilon[u \geq \lambda + \frac{1}{n}] = \bigcup \{O \in \mathcal{CC}([u > \lambda]), |O| > \varepsilon\}. \tag{3.3}$$

If $C \in \mathcal{CC}([u \geq \lambda + \frac{1}{n}])$ and $|C| \geq \varepsilon$ for some n, we define $O = \mathrm{cc}([u > \lambda], C)$. Since O is open and contains the closed set C, we deduce that $O \setminus C$ is open and not empty, and, thus, of positive measure. Hence $|O| > |C| \geq \varepsilon$. This proves that

$$\bigcup_{n>0} M_\varepsilon([u \geq \lambda + \frac{1}{n}]) \subseteq \bigcup \{O \in \mathcal{CC}([u > \lambda]), |O| > \varepsilon\}.$$

Now, let $O = \mathrm{cc}([u > \lambda], x)$ be such that $|O| > \varepsilon$. Then, thanks to Lemma 3.2, we have

$$\mathrm{cc}([u > \lambda], x) = \bigcup_{n>0} \mathrm{cc}([u \geq \lambda + \frac{1}{n}], x).$$

Thus, $\varepsilon < |O| = \sup_{n>0} |\mathrm{cc}([u \geq \lambda + \frac{1}{n}], x)|$. There is some n such that if $C = \mathrm{cc}([u \geq \lambda + \frac{1}{n}], x)$, then $|C| > \varepsilon$. This proves the remaining inclusion in (3.3). The right hand side of this equality being open, we conclude that $[M_\varepsilon^+ u \le \lambda]$ is closed.

To prove the same result for M_ε^-, we write

$$[M_\varepsilon^- u \le \lambda] = \bigcap_{n>0} [M_\varepsilon^- u < \lambda + \frac{1}{n}]$$
$$= \bigcap_{n>0} M'_\varepsilon([u < \lambda + \frac{1}{n}]).$$

We claim that the last set coincides with $\bigcup\{C \in \mathcal{CC}([u \le \lambda]), |C| \ge \varepsilon\}$. If $C \in \mathcal{CC}([u \le \lambda])$ and $|C| \ge \varepsilon$, we define $O_n = \mathrm{cc}([u < \lambda + \frac{1}{n}], C)$. Since O_n is open and C is closed, we have that $\varepsilon \le |C| < |O_n|$. Hence,

$$\bigcup\{C \in \mathcal{CC}([u \le \lambda]), |C| \ge \varepsilon\} \subseteq \bigcap_{n>0} M'_\varepsilon([u < \lambda + \frac{1}{n}]).$$

Let $x \in M'_\varepsilon([u < \lambda + \frac{1}{n}])$ for all n. Then we have

$$\mathrm{cc}([u \le \lambda], x) \subseteq \bigcap_{n>0} \mathrm{cc}([u < \lambda + \frac{1}{n}], x) \subseteq \bigcap_{n>0} \mathrm{cc}([u \le \lambda + \frac{1}{n}], x) = \mathrm{cc}([u \le \lambda], x),$$

where the last equality is justified by Lemma 3.2. Hence $|\mathrm{cc}([u \le \lambda], x)| = \inf_n |\mathrm{cc}([u < \lambda + \frac{1}{n}], x)| \ge \varepsilon$. This proves that

$$\bigcap_{n>0} M'_\varepsilon([u < \lambda + \frac{1}{n}]) \subseteq \bigcup\{C \in \mathcal{CC}([u \le \lambda]), |C| \ge \varepsilon\},$$

and, thus, the equality of both terms. The right hand side term, being a finite union of closed sets, it is also closed. □

Proposition 3.9. *When restricted to upper semicontinuous functions, both M_ε^+ and M_ε^- are idempotent.*

Proof. Let u be an upper semicontinuous function and $\lambda \in \mathbb{R}$. Clearly, we have $M_\varepsilon \circ M_\varepsilon = M_\varepsilon$. Applying this equality to the set $[u \ge \lambda]$, and using Corollary 3.5, we get

$$M_\varepsilon([M_\varepsilon^+ u \ge \lambda]) = [M_\varepsilon^+ u \ge \lambda].$$

Now, thanks to Proposition 3.8, we have that $[M_\varepsilon^+ u \ge \lambda]$ is closed and we can apply again Corollary 3.5 to the left hand side of this equality to obtain

$$[M_\varepsilon^+ M_\varepsilon^+ u \ge \lambda] = [M_\varepsilon^+ u \ge \lambda].$$

Since this equality holds for any $\lambda \in \mathbb{R}$, we conclude that $M_\varepsilon^+ \circ M_\varepsilon^+ = M_\varepsilon^+$.

The same proof applies to M_ε^-. □

Proposition 3.10. *Let* $u \in C(\overline{\Omega})$*. Then*

(i) $M_\varepsilon^+ u \uparrow u$, $M_\varepsilon^- u \downarrow u$ *as* $\varepsilon \to 0+$.
(ii) $M_\varepsilon^+ M_\varepsilon^- u \to u$, $M_\varepsilon^- M_\varepsilon^+ u \to u$ *uniformly as* $\varepsilon \to 0+$.

Proof. (i) Both cases being similar, it will be sufficient to prove the first part of the assertion. Let $v(x) = \sup_\varepsilon M_\varepsilon^+ u(x)$. Observe that v is lower semicontinuous. Assume that $M := \max_{x \in \overline{\Omega}}(u(x) - v(x)) > 0$. Let $x_0 \in \overline{\Omega}$ be such that $M = u(x_0) - v(x_0)$. Let $C = cc([u \geq v(x_0) + \frac{M}{2}], x_0)$. Since u is continuous, C contains an open set, hence, $|C| > \delta > 0$ for some δ. Let $\varepsilon \leq \delta$. Then

$$M_\varepsilon^+ u(x) \geq \inf_{y \in C} u(y) \geq v(x_0) + \frac{M}{2} \quad \text{for all } x \in C.$$

Letting $\varepsilon \to 0+$ and choosing $x = x_0$ we obtain

$$v(x_0) \geq v(x_0) + \frac{M}{2}$$

which is a contradiction. We conclude that $v(x) = u(x)$ and the proposition is proved.
(ii) Both cases being similar, we shall only prove that $M_\varepsilon^+ M_\varepsilon^- u \to u$ uniformly as $\varepsilon \to 0+$. For that, let us write

$$M_\varepsilon^+ M_\varepsilon^- u - u = M_\varepsilon^+ M_\varepsilon^- u - M_\varepsilon^+ u + M_\varepsilon^+ u - u.$$

Given $\delta > 0$, let $\varepsilon_0 > 0$ be small enough so that

$$u - \delta \leq M_\varepsilon^- u \leq u + \delta$$

and

$$|M_\varepsilon^+ u - u| \leq \delta$$

for all $\varepsilon \leq \varepsilon_0$. Then

$$M_\varepsilon^+ u - \delta \leq M_\varepsilon^+ M_\varepsilon^- u \leq M_\varepsilon^+ u + \delta,$$

i.e.,

$$|M_\varepsilon^+ M_\varepsilon^- u - M_\varepsilon^+ u| \leq \delta.$$

Collecting these facts we have that

$$|M_\varepsilon^+ M_\varepsilon^- u - u| \leq 2\delta,$$

for all $\varepsilon \leq \varepsilon_0$. □

Remark. If $A : \mathbb{R}^N \to \mathbb{R}^N$ is a linear map whose determinant we denote by *det* A, if $u : \mathbb{R}^N \to \mathbb{R}^N$ is a measurable map and $Au(x) = u(Ax), x \in \mathbb{R}^N$, then $M_\varepsilon^+ Au(x) = M_{|det\,A|\varepsilon}^+ u(Ax)$ and $M_\varepsilon^- Au(x) = M_{|det\,A|\varepsilon}^- u(Ax)$, for all $x \in \mathbb{R}^N$. In particular, the area opening and closing are invariant under special affine maps, i.e., linear maps with $|det\,A| = 1$.

3.3.4 Interpretation

Let u be an upper semicontinuous function and $C \in \mathcal{CC}([u \geq \lambda])$. If $|C| \geq \varepsilon$, we have $M_\varepsilon(C) = C$ and, thus, C is a connected component of $M_\varepsilon([u \geq \lambda]) = [M_\varepsilon^+ u \geq \lambda]$. If $|C| < \varepsilon$, then $M_\varepsilon(C) = \emptyset$ and C is *not* a connected component of $[M_\varepsilon^+ u \geq \lambda]$: it does not even meet this set, since $[M_\varepsilon^+ u \geq \lambda] = M_\varepsilon([u \geq \lambda])$, so that

$$[M_\varepsilon^+ u \geq \lambda] \cap C = M_\varepsilon([u \geq \lambda]) \cap C = \emptyset.$$

Conversely, if $C \in \mathcal{CC}([M_\varepsilon^+ u \geq \lambda])$, we have $C \in \mathcal{CC}(M_\varepsilon([u \geq \lambda]))$ and $C = M_\varepsilon(C')$, C' being a connected component of $[u \geq \lambda]$, and since $C \neq \emptyset$, we have $M_\varepsilon(C') = C'$, thus $C = C'$.

Summing up these remarks, we can see that the connected components of $[M_\varepsilon^+ u \geq \lambda]$ are exactly the connected components of $[u \geq \lambda]$ of measure $\geq \varepsilon$. In particular, since the connected components of upper level sets have a structure of tree driven by inclusion, the tree of $M_\varepsilon^+ u$ is the tree of u pruned of all nodes of insufficient measure.

The same observations can be made concerning M_ε^- and the connected components of $[u < \lambda]$. The tree of M_ε^- is the tree of connected components of lower level sets of u pruned of all nodes of insufficient measure.

Summarizing the above discussion, we have the following result.

Proposition 3.11. *If $X \in \mathcal{CC}([M_\varepsilon^- u < \lambda]) \neq \emptyset$, then $|X| > \varepsilon$. If $X \in \mathcal{CC}([M_\varepsilon^+ u \geq \lambda]) \neq \emptyset$, then $|X| \geq \varepsilon$. If $|cc([u \geq \lambda], x)| \geq \varepsilon$ (resp. if $|cc([u < \lambda], x)| \geq \varepsilon$) then $|cc([M_\varepsilon^- u \geq \lambda], x)| \geq \varepsilon$ (resp. $|cc([M_\varepsilon^+ u < \lambda], x)| \geq \varepsilon$).*

In particular, the above result implies that the connected components of $[M_\varepsilon^+ M_\varepsilon^- u \geq \lambda]$ and $[M_\varepsilon^+ M_\varepsilon^- u < \lambda]$ have measure $\geq \varepsilon$. The same thing can be said of connected components of the upper and lower level sets of $M_\varepsilon^- M_\varepsilon^+ u$. For that reason, the operators $M_\varepsilon^+ M_\varepsilon^-$ and $M_\varepsilon^- M_\varepsilon^+ u$ are called extrema filters.

Let us finally mention that, when defined on $\mathbb{R}^N$ these operators are translation invariant, since the notions of connected set and measure are translation invariant.

3.3.5 Composition

Lemma 3.12. *Let u be an upper semicontinuous function and $\lambda \in \mathbb{R}$. Then*

$$C \in \mathcal{CC}([M_\varepsilon^+ u < \lambda]), C \neq \emptyset \Rightarrow \exists x \in C, \quad u(x) < \lambda.$$
$$C \in \mathcal{CC}([M_\varepsilon^- u \geq \lambda]), C \neq \emptyset \Rightarrow \exists x \in C, \quad u(x) \geq \lambda.$$

Proof. Let $C \in \mathcal{CC}([M_\varepsilon^+ u < \lambda])$. If the consequence was false, we would have $C \subseteq [u \geq \lambda]$. If $C = \overline{\Omega}$, then we have

$$[M_\varepsilon^+ u \geq \lambda] = M_\varepsilon([u \geq \lambda]) = M_\varepsilon \overline{\Omega} = \overline{\Omega},$$

since $|\overline{\Omega}| \geq \varepsilon$. This contradicts the definition of C. If $C \subsetneq \overline{\Omega}$, C being open, we have $\overline{C} \supsetneq C$ and $\partial C \subseteq [M_\varepsilon^+ u \geq \lambda] = M_\varepsilon([u \geq \lambda])$. Let $x \in \partial C$. Then x belongs to a connected component D of $[u \geq \lambda]$ such that $|D| \geq \varepsilon$. Since $x \in \overline{C}$, $D \cup \overline{C}$ is connected and included in $[u \geq \lambda]$. As $|D \cup \overline{C}| \geq \varepsilon$, we get $D \cup \overline{C} \subseteq [M_\varepsilon^+ u \geq \lambda]$, contradicting the definition of C. This proves our claim that there is some $x \in C$ such that $u(x) < \lambda$.

Now, let $C \in \mathcal{CC}([M_\varepsilon^- u \geq \lambda])$. Suppose that $C \subseteq [u < \lambda]$. If $C = \overline{\Omega}$, arguing as above, we have $[M_\varepsilon^- u < \lambda] = \overline{\Omega}$, which contradicts the hypothesis. Thus, we may assume that $C \subsetneq \overline{\Omega}$. Let $D = \mathrm{cc}([u < \lambda], C)$. Since C is closed and D open, then $C \subsetneq D$. Thus there is some x in $D \setminus C$ such that $M_\varepsilon^- u(x) < \lambda$. Thus, $x \in M_\varepsilon'([u < \lambda])$, meaning that $|\,\mathrm{cc}([u < \lambda], x)| > \varepsilon$. This component must be D and, therefore $|D| > \varepsilon$. Finally, we observe that $D \subseteq [M_\varepsilon^- u < \lambda]$, which is a contradiction. □

Theorem 3.13. *The operators* $M_\varepsilon^+ \circ M_\varepsilon^-$ *and* $M_\varepsilon^- \circ M_\varepsilon^+$

(i) transform upper semicontinuous functions into upper semicontinuous functions, and continuous functions into continuous functions;
(ii) are idempotent on upper semicontinuous functions.

Proof. (i) is a direct consequence of the equivalent properties we have proved for M_ε^+ and M_ε^- in Proposition 3.8.

(ii) is a consequence of the well-known result in mathematical morphology that the composition of a morphological opening and closing are idempotent. We shall include a proof for the sake of completeness. As a consequence of Proposition 3.6, we have $M_\varepsilon^+ u \leq u \leq M_\varepsilon^- u$ for any function u. Applying this to $M_\varepsilon^+ M_\varepsilon^- u$ instead of u, we get

$$M_\varepsilon^+ M_\varepsilon^- u \leq M_\varepsilon^- M_\varepsilon^+ M_\varepsilon^- u.$$

We apply the first part of Lemma 3.12 to $v = M_\varepsilon^- u$, $\lambda \in \mathbb{R}$ and $C \in \mathcal{CC}([M_\varepsilon^+ v < \lambda])$ such that $C \neq \emptyset$. We find a point $x \in C \cap [v < \lambda]$. Let $D = \mathrm{cc}([v < \lambda], x)$. We know that $|D| \geq \varepsilon$. Since $M_\varepsilon^+ v \leq v$, we have that $D \subseteq [M_\varepsilon^+ v < \lambda]$. Since $D \cap C \neq \emptyset$ and D is connected, we have $D \subseteq C$ and, therefore, $|C| \geq \varepsilon$. Thus, $M_\varepsilon'(C) = C$, yielding

$$C \subseteq [M_\varepsilon^- M_\varepsilon^+ M_\varepsilon^- u < \lambda].$$

This proves that $M_\varepsilon^- M_\varepsilon^+ M_\varepsilon^- u \leq M_\varepsilon^+ M_\varepsilon^- u$, and, thus, we have the equality $M_\varepsilon^- M_\varepsilon^+ M_\varepsilon^- u = M_\varepsilon^+ M_\varepsilon^- u$. Applying M_ε^+ to each member and using its idempotency, we conclude that

$$M_\varepsilon^+ M_\varepsilon^- M_\varepsilon^+ M_\varepsilon^- u = M_\varepsilon^+ M_\varepsilon^- u.$$

With a similar proof, using the second part of Lemma 3.12, we prove that $M_\varepsilon^- M_\varepsilon^+$ is idempotent. □

3.3.6 The Effect of Extrema Filters on the Tree of Shapes

Proposition 3.14. *If A is a closed set, C a connected component of A and H an internal hole of C, then, for any x in H,*

$$H = \bigcup_{G=\mathrm{Sat}(G'):G'\in\mathcal{CC}(\overline{\Omega}\setminus A),\, x\in G\subseteq H} G.$$

Thus, any hole of C can be expressed as a countable union of saturations of connected components of $\overline{\Omega} \setminus A$.

Proof. Let H' denote the right hand side of the above equality. Observe that, by Lemma 2.12, $H' \neq \emptyset$. Obviously, $H' \subseteq H$. Suppose that this inclusion is strict. The set H being connected and H' being open (as a union of open sets), H' is not closed in H, so that $H \cap \partial H' \neq \emptyset$. Let $y \in H \cap \partial H'$ and let U be a connected neighborhood of y which we may take contained in H, since H is an open set. Observe that $U \cap H' \neq \emptyset$, $U \cap (\overline{\Omega} \setminus H') \neq \emptyset$. In particular, there is $G' \in \mathcal{CC}(\overline{\Omega} \setminus A)$ such that $G = \mathrm{Sat}(G')$ satisfies $x \in G \subseteq H'$ and $U \cap G \neq \emptyset$. We claim that $U \cap A \neq \emptyset$. Otherwise, we would have $U \subseteq \overline{\Omega} \setminus A$. In that case $G' \subseteq U \cup G' \subseteq \overline{\Omega} \setminus A$. Since $G' \subseteq H'$ and $U \cap (\overline{\Omega} \setminus H') \neq \emptyset$, U cannot be contained in G' and, thus, $G' \subsetneq G' \cup U$. Let us observe that $U \cap G' \neq \emptyset$ Indeed, $U \cap G \neq \emptyset$ and $y \in U \setminus H' \subseteq U \setminus G$. Thus $U \cap \partial G \neq \emptyset$. Since $\partial G = \partial\,\mathrm{Sat}(G') \subseteq \partial G'$, we have that $U \cap \partial G' \neq \emptyset$, and, therefore, also $U \cap G' \neq \emptyset$. It follows that $U \cup G' \subseteq \overline{\Omega} \setminus A$ and strictly contains G'. This contradiction proves that $U \cap A \neq \emptyset$. We conclude that $y \in \overline{A} = A$. Since H is a hole of $C \in \mathcal{CC}(A)$ and $y \in H$, we may apply Lemma 2.12, and find $O' \in \mathcal{CC}(\overline{\Omega} \setminus A)$ such that $y \in \mathrm{Sat}(O') \subseteq H \subseteq \mathrm{Sat}(C)$. Let $O = \mathrm{Sat}(O')$, O, O' being open sets. Let V be a neighborhood of y such that $V \subseteq O$, and, since $y \in \partial H'$, V meets H' and $\overline{\Omega} \setminus H'$. Since $V \cap H' \neq \emptyset$, there is $Q' \in \mathcal{CC}(\overline{\Omega} \setminus A)$ such that $Q = \mathrm{Sat}(Q')$ satisfies $x \in Q \subseteq H'$ and such that $V \cap Q \neq \emptyset$. Thus, also $O \cap Q \neq \emptyset$. Both, O and Q being shapes, they must be nested. If $O \subseteq Q$, then $V \subseteq Q \subseteq H'$ and V could not intersect $\overline{\Omega} \setminus H'$, a contradiction. We must have $Q \subseteq O$. Since $x \in Q$, also $x \in O \subseteq H$. We conclude that $O \subseteq H'$. Then $y \in H'$, a contradiction. We conclude that $H = H'$. □

Lemma 3.15. *Let A be a closed subset of $\overline{\Omega}$, and let Y be a connected component of $\overline{\Omega} \setminus A$ such that $p_\infty \notin Y$. Then* $\mathrm{Sat}(Y)$ *is a hole of a connected component of A.*

Proof. Since Y is open, also $\mathrm{Sat}(Y)$ is open. Hence $\partial\,\mathrm{Sat}(Y) \subseteq A$. Let $Z = cc(A, \partial\,\mathrm{Sat}(Y))$. We have

$$Y \subseteq \mathrm{Sat}(Y) \subseteq \mathrm{Sat}(\partial\,\mathrm{Sat}(Y)) \subseteq \mathrm{Sat}(Z).$$

Since $Y \cap Z = \emptyset$, then Y is contained in an internal hole H of Z. Thus, also $\mathrm{Sat}(Y) \subseteq H$. If $\mathrm{Sat}(Y) \neq H$, then, as in the proof of Proposition 3.14, we would have that $\partial\,\mathrm{Sat}(Y) \cap H \neq \emptyset$. Hence $Z \cap H \neq \emptyset$. This contradiction proves that $\mathrm{Sat}(Y) = H$. □

For simplicity, we shall say that X is a limit shape of lower (upper) type of u if X is the union (resp. intersection) of a family of nested shapes of u of lower (resp. upper) type.

Theorem 3.16. *Let $u : \overline{\Omega} \to \mathbb{R}$ be an upper semicontinuous function. We have*

(i) Assume that u is bounded. If X is a shape of upper type of $M_\varepsilon^+ u$ (resp. of $M_\varepsilon^- u$), then X is a shape of upper type of u.
(ii) If X is a shape of lower type of $M_\varepsilon^+ u$ (resp. of $M_\varepsilon^- u$), then X is a limit shape (resp. a shape) of lower type of u.

Proof. *(i)* Let X be a shape of upper type of $M_\varepsilon^+ u$. Let $\lambda \in \mathbb{R}$, $Y \in \mathcal{CC}([M_\varepsilon^+ u \geq \lambda])$ be such that $X = \mathrm{Sat}(Y)$. Since $[M_\varepsilon^+ u \geq \lambda] = M_\varepsilon^+[u \geq \lambda]$, then $Y \in \mathcal{CC}([u \geq \lambda])$. Thus X is a shape of upper type of u.

Let X be a shape of upper type of $M_\varepsilon^- u$. Let $\lambda \in \mathbb{R}$, $Y \in \mathcal{CC}([M_\varepsilon^- u \geq \lambda])$ be such that $X = \mathrm{Sat}(Y)$. Notice that

$$[M_\varepsilon^- u \geq \lambda] = \overline{\Omega} \setminus [M_\varepsilon^- u < \lambda] = \overline{\Omega} \setminus M_\varepsilon'[u < \lambda] = \overline{\Omega} \setminus \bigcup_{Z \in \mathcal{CC}([u<\lambda]), |Z| > \varepsilon} Z$$

$$= [u \geq \lambda] \cup \bigcup_{Z \in \mathcal{CC}([u<\lambda]), |Z| \leq \varepsilon} Z$$

Let us consider first the case where $X = \overline{\Omega}$. Since u is bounded, we have that $\overline{\Omega}$ is also a shape of upper type of u. Thus, we may assume that $X \neq \overline{\Omega}$, hence $p_\infty \notin Y$. Then, by Lemma 2.10, we have that $\mathrm{Sat}(Y) = \mathrm{Sat}(\partial Y)$. Let us admit for a moment that $\partial Y \subseteq [u \geq \lambda]$, and finish the proof. Let $Z = cc([u \geq \lambda], \partial Y)$. Observe that $\partial Y \subseteq Z \cap Y$, hence $\partial Y \subseteq Z \subseteq Y$. Thus

$$\mathrm{Sat}(\partial Y) \subseteq \mathrm{Sat}(Z) \subseteq \mathrm{Sat}(Y).$$

We conclude that $\mathrm{Sat}(Y) = \mathrm{Sat}(Z)$, and, thus, X is a shape of upper type of u.

Let us prove that $\partial Y \subseteq [u \geq \lambda]$. Otherwise, if $p \in \partial Y$, $u(p) < \lambda$, then $B(p,r) \subseteq [u < \lambda]$ for some $r > 0$. Hence $B(p,r)$ is contained in a connected component of $[u < \lambda]$. Since $p \in Y$, by the above formula for $[M_\varepsilon^- u \geq \lambda]$, we would have that $B(p,r) \subseteq Y$. Hence p cannot be in ∂Y. This contradiction proves our assertion.
(ii) Let X be a shape of lower type of $M_\varepsilon^+ u$. Let $\lambda \in \mathbb{R}$, $Y \in \mathcal{CC}([M_\varepsilon^+ u < \lambda])$ be such that $X = \mathrm{Sat}(Y)$. Again we observe that, if $X = \overline{\Omega}$, and u being upper bounded, we have that X is also a shape of lower type of u. Thus, we may assume that $X = \mathrm{Sat}(Y) \neq \overline{\Omega}$, and $p_\infty \notin Y$. Recall that

$$\overline{\Omega} \setminus [M_\varepsilon^+ u < \lambda] = \bigcup_{Z \in \mathcal{CC}([u \geq \lambda]), |Z| \geq \varepsilon} Z.$$

Now, we use Lemma 3.15 with A being the set in the above equality, and Y the set we are considering here, we conclude that $\mathrm{Sat}(Y)$ is a hole of a connected component of A. Since connected components of A are connected components of $[u \geq \lambda]$, we have that $\mathrm{Sat}(Y)$ is a hole of a connected component of $[u \geq \lambda]$, call it C. Using Proposition 3.14, we conclude that

$$\mathrm{Sat}(Y) = \bigcup_{G \in \mathcal{CC}([u<\lambda]), x \in \mathrm{Sat}(G) \subseteq \mathrm{Sat}(Y)} \mathrm{Sat}(G)$$

where x is any point in $\mathrm{Sat}(Y)$. Since, any tho shapes are either nested or disjoint, we conclude that the above union can be expressed as a countable increasing union. Therefore $\mathrm{Sat}(Y)$ is a limit shape.

Let X be a shape of lower type of $M_\varepsilon^- u$. Let $\lambda \in \mathbb{R}$, $Y \in \mathcal{CC}([M_\varepsilon^+ u < \lambda])$ be such that $X = \mathrm{Sat}(Y)$. Since $[M_\varepsilon^+ u < \lambda] = M_\varepsilon'[u < \lambda]$, then $Y \in \mathcal{CC}([u < \lambda])$. Thus X is a shape of lower type of u. □

Lemma 3.17. *Let $u : \overline{\Omega} \to \mathbb{R}$ be an upper semicontinuous function. Let $\mathrm{Sat}(A_n)$, $\mathrm{Sat}(B_n)$ be increasing sequences of shapes of u. If $\cup_n \mathrm{Sat}(A_n) \subsetneq \cup_n \mathrm{Sat}(B_n)$, then there is $k \in \mathbb{N}$ such that $\cup_n \mathrm{Sat}(A_n) \subseteq \mathrm{Sat}(B_k)$.*

Proof. We have to prove

$$\exists k \in \mathbb{N} \text{ such that } \forall n \quad \mathrm{Sat}(A_n) \subseteq \mathrm{Sat}(B_k).$$

Otherwise

$$\forall k \in \mathbb{N}\ \exists n_k \text{ such that } \mathrm{Sat}(A_{n_k}) \not\subseteq \mathrm{Sat}(B_k). \tag{3.4}$$

Since $\mathrm{Sat}(A_n)$ are increasing, (3.4) is equivalent to

$$\forall k \in \mathbb{N}\ \exists n_k \text{ such that } \mathrm{Sat}(A_n) \not\subseteq \mathrm{Sat}(B_k) \quad \forall n \geq n_k. \tag{3.5}$$

Since any two shapes of u are nested or disjoint, we may write

$$\forall k \in \mathbb{N}\ \exists n_k \text{ such that if } n \geq n_k \text{ then}$$
$$\text{either } \mathrm{Sat}(A_n) \cap \mathrm{Sat}(B_k) = \emptyset \text{ or } \mathrm{Sat}(B_k) \subseteq \mathrm{Sat}(A_n). \tag{3.6}$$

Let $I := \{k : \exists m_k \geq n_k \text{ s.t. } \mathrm{Sat}(B_k) \subseteq \mathrm{Sat}(A_{m_k})\}$. If I is an infinite set, then we would obtain $\cup_n \mathrm{Sat}(B_n) \subseteq \cup_n \mathrm{Sat}(A_n)$, a contradiction with our assumptions. Hence, the set I is finite. Let k' be its maximum element. Then

$$\forall k \geq k'\ \exists n_k \text{ such that } \mathrm{Sat}(A_n) \cap \mathrm{Sat}(B_k) = \emptyset \quad \forall n \geq n_k.$$

This implies that $\left(\cup_n \mathrm{Sat}(B_n)\right) \cap \left(\cup_n \mathrm{Sat}(A_n)\right) = \emptyset$, again a contradiction. The Lemma follows. □

Proposition 3.18. *Let $u : \overline{\Omega} \to \mathbb{R}$ be a bounded upper semicontinuous function. If X is a limit shape of $M_\varepsilon^+ u$ (resp. of $M_\varepsilon^- u$), then X is a limit shape of u. In other words, the tree structure of $M_\varepsilon^+ u$ (resp. $M_\varepsilon^- u$) is a simplified version of the tree structure of u.*

A similar statement can be done for the tree structure of the functions $M_\varepsilon^+ M_\varepsilon^- u$ and $M_\varepsilon^- M_\varepsilon^+ u$.

Proof. Let X be a limit shape of $M_\varepsilon^+ u$. By Proposition 2.23 we know that we may write X in one of the following forms:

(i) $X = \cup_n X_n$, where X_n is a shape of upper type of $M_\varepsilon^+ u$.
(ii) $X = \cup_n X_n$, where X_n is a shape of lower type of $M_\varepsilon^+ u$.
(iii) $X = \cap_n X_n$, where X_n is a shape of upper type of $M_\varepsilon^+ u$.
(iv) $X = \cap_n X_n$, where X_n is a shape of lower type of $M_\varepsilon^+ u$.

Assume that we are in case (i). Then, by Theorem 3.16, X_n are also shapes of upper type of u, hence X is a limit shape of u.

Assume that we are in case (ii). By Theorem 3.16, if X_n is the saturation of a connected component of $[M_\varepsilon^+ u < \lambda_n]$, $\lambda_n \in \mathbb{R}$, then we may write $X_n = \cup_m \mathrm{Sat}(X_{nm})$ where $X_{nm} \in \mathcal{CC}([u < \lambda_n])$. Then $X = \cup_n \cup_m \mathrm{Sat}(X_{nm})$. Since we may reorder the sets $\mathrm{Sat}(X_{nm})$ in increasing order, we see that X is a limit shape of u.

Assume that we are in case (iii). Then, by Theorem 3.16, X_n are also shapes of upper type of u, hence X is a limit shape of u.

Assume that we are in case (iv). Without loss of generality we may assume that $X_n \supsetneq X_{n+1}$. By Theorem 3.16, if X_n is the saturation of a connected component of $[M_\varepsilon^+ u < \lambda_n]$, $\lambda_n \in \mathbb{R}$, then we may write $X_n = \cup_m \mathrm{Sat}(X_{nm})$ where $X_{nm} \in \mathcal{CC}([u < \lambda_n])$. Then $X = \cap_n \cup_m \mathrm{Sat}(X_{nm})$. For each $n \geq 1$, let m_n be the integer given by Lemma 3.17 such that

$$\cup_m \mathrm{Sat}(X_{(n+1)m}) \subseteq \mathrm{Sat}(X_{n,m_n}).$$

Then we have

$$\begin{aligned} X &= \cap_{n\geq 1} \cup_m \mathrm{Sat}(X_{nm}) = \\ &= \left(\cup_m \mathrm{Sat}(X_{1m})\right) \cap \left(\cup_m \mathrm{Sat}(X_{2m})\right) \cap \left(\cap_{n\geq 3} \cup_m \mathrm{Sat}(X_{nm})\right) \\ &= \mathrm{Sat}(X_{1m_1}) \cap \left(\cup_m \mathrm{Sat}(X_{2m})\right) \cap \left(\cap_{n\geq 3} \cup_m \mathrm{Sat}(X_{nm})\right) \\ &= \dots \\ &= \cap_{i=1}^r \mathrm{Sat}(X_{im_i}) \cap \left(\cap_{n\geq r+1} \cup_m \mathrm{Sat}(X_{nm})\right) \\ &= \cap_{i=1}^\infty \mathrm{Sat}(X_{im_i}). \end{aligned}$$

We conclude that X is a limit shape of u.

The proof is similar when X is a limit shape of $M_\varepsilon^- u$. □

As a consequence of Proposition 4.26 (see Chap. 4), if $u \in C(\overline{\Omega})$ is such that $M_\varepsilon^- u = M_\varepsilon^+ u = u$, then all limit shapes of u are shapes. This result is not true if $u : \overline{\Omega} \to \mathbb{R}$ is an upper semicontinuous function such that $M_\varepsilon^- u = M_\varepsilon^+ u = u$. Indeed, it suffices to consider $u(x) = 0$ for $x \in B(0,1) \setminus (-1,1) \times \{0\}$, $u(x) = 1$ for $x \notin B(0,1)$, $u(x)$ decreasing from 1 to $a > 0$ when $x \in (-1,0] \times \{0\}$, and $u(x)$ decreasing from 1 to 0 when $x \in (0,1) \times \{0\}$. Let p be any point in $B(0,1) \setminus (-1,1) \times \{0\}$. Then the limit shape $\cap_n \mathrm{Sat}(cc([u < \frac{1}{n}], p)) = \cap_n cc([u < \frac{1}{n}], p) = B(0,1) \setminus (-1,1) \times \{0\}$ is not a shape.

3.4 Grain Filter

3.4.1 Introduction and Definitions

The area opening, area closing and the extrema filters studied above are connected operators (see Sect. 3.5.2). We have seen that M_ε^- and M_ε^+ do not commute, and neither M_ε^-, M_ε^+, $M_\varepsilon^- M_\varepsilon^+$ nor $M_\varepsilon^+ M_\varepsilon^-$ deals symmetrically with upper and lower level sets. Actually, they work in two steps: first upper, then lower level sets are treated (or in the opposite order). These operators are not self-dual.

We are going to introduce now an operator which is self-dual when acting on continuous functions. We shall call it the grain filter and denote it by G_ε . First we introduce the grain filter for sets and then we lift it to functions by the usual procedure described in Sect. 3.2.2. Recall that, besides the monotonicity of the operator on sets, a necessary and sufficient condition for that extension to be possible is the upper semicontinuity of the operator on sets, i.e,

$$G_\varepsilon(\bigcap_n F_n) = \bigcap_n G_\varepsilon(F_n) \tag{3.7}$$

holds for any decreasing sequence of sets F_n. Indeed, if we want to extend G_ε to upper semicontinuous functions it suffices that the above property holds for any decreasing sequence of closed sets. We shall check that this is indeed true. Now, the self-duality on functions would be implied by the self-duality on sets (see below and Sect. 3.5.2 for definitions). The problem is that the grain filter G_ε as we shall define it is not self-dual on sets. In spite of this, it is self-dual on functions, the reason being that it is self-dual for sets as far as the sets satisfy some additional conditions. And these additional conditions hold for almost all level sets of a continuous function. Thus the set operator G_ε is self-dual for almost all level sets of a continuous function and this is sufficient to lift the self-duality to the extended operator $\tilde{G}_\varepsilon$ when acting on continuous functions. Since the above strategy is quite involved, we shall directly prove the self-duality of $\tilde{G}_\varepsilon$ on continuous functions; later, in Sect. 3.5.2, having all details at hand, we shall give more details on the above strategy.

Definition 3.19. A contrast invariant operator $\tilde{T}$, associated to the set operator T, is said to be self-dual on continuous functions if the following equivalent properties hold for any continuous function u

1. $\tilde{T}(-u) = -\tilde{T}(u)$.
2. $\forall x$, $\sup_{B\in\mathcal{B}} \inf_{y\in B} u(y) = \inf_{B\in\mathcal{B}} \sup_{y\in B} u(y)$, $\mathcal{B}$ describing the structuring elements of $\tilde{T}$.
3. $\forall\lambda$, $T([u \leq \lambda]) = [\tilde{T}u \leq \lambda]$.

To show the equivalence of the first two properties, it suffices to write the second one with $-u$ instead of u. By taking $-\lambda$ instead of λ, we can see that the third property is equivalent to

$$\forall\lambda, \quad T([-u \geq \lambda]) = [-\tilde{T}u \geq \lambda]. \tag{3.8}$$

Now, using the contrast invariance of $\tilde{T}$, (3.8) amounts to write

$$\forall\lambda, \quad [\tilde{T}(-u) \geq \lambda] = [-\tilde{T}u \geq \lambda];$$

which means exactly that $\tilde{T}(-u) = -\tilde{T}u$.

Since an area threshold is involved, we define our set operator on measurable sets. For the developments of this section it would be sufficient to define it on closed sets, but for later discussion we define it in the general context. If X is a measurable set, we define

$$\begin{aligned} G_\varepsilon X = \bigcup \{ \operatorname{Sat}(C) \setminus \cup_i C'_i : \; & C \in \mathcal{CC}(X),\, |\operatorname{Sat}(C)| \geq \varepsilon, \\ & C'_i \text{ internal hole of } C,\, |C'_i| > \varepsilon\}. \end{aligned} \tag{3.9}$$

Since, if $|\operatorname{Sat}(C)| \geq \varepsilon$, then

$$G_\varepsilon(C) = \operatorname{Sat}(C) \setminus \{\cup_i C'_i : C'_i \text{ internal hole of } C,\, |C'_i| > \varepsilon\},$$

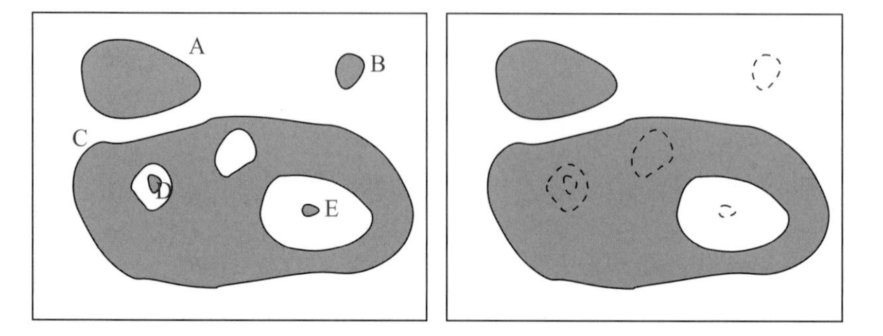

Fig. 3.1 Application of operator defined in Formula (3.9) to a set X composed of 5 connected components, A to E. B, D and E are too small, they are removed. A is sufficiently large and has no hole, it remains unchanged. Two holes of C are too small, they are filled, whereas the hole containing E is large enough, and remains a hole.

we may write

$$G_\varepsilon X = \bigcup\{G_\varepsilon C : C \in \mathcal{CC}(X), |\operatorname{Sat}(C)| \geq \varepsilon\}. \tag{3.10}$$

An example of such an operation applied to a set is illustrated in Fig. 3.1. It shows that connected components are processed independently. For $\varepsilon > |\overline{\Omega}|$, we clearly have that $G_\varepsilon X = \emptyset$ for any measurable set X. Thus we will always suppose that $\varepsilon \leq |\overline{\Omega}|$.

Observe that

$$\begin{aligned} G_\varepsilon X = \bigcup \{\, C \cup \cup_i C_i' : \, & C \in \mathcal{CC}(X), |\operatorname{Sat}(C)| \geq \varepsilon, \\ & C_i' \text{ internal hole of } C, |C_i'| \leq \varepsilon\}. \end{aligned} \tag{3.11}$$

3.4.2 Preliminary Results

Let $X \subseteq \overline{\Omega}$. For simplicity, we shall denote by $\mathcal{H}_c(X)$ the family of the internal holes of the connected components of X.

Lemma 3.20. *G_ε is nondecreasing on measurable sets.*

Proof. Let $X \subseteq Y$ be measurable sets. Let $C \in \mathcal{CC}(X)$ be such that $|\operatorname{Sat}(C)| \geq \varepsilon$. Then there exists $C' \in \mathcal{CC}(Y)$ such that $C \subseteq C'$. On the other hand, if H' is an internal hole of C' such that $|H'| > \varepsilon$, then H' is included in an hole H of C, and, thus, $|H| > \varepsilon$. It could happen that H is not an internal hole, in that case, $H = H_\infty := \overline{\Omega} \setminus \operatorname{Sat}(C)$ Thus

$$\bigcup\{H' : H' \in \mathcal{H}_c(C'), |H'| > \varepsilon\} \subseteq \bigcup\{H : H \in \mathcal{H}_c(C), |H| > \varepsilon\} \cup H_\infty$$

and we deduce that

$$\begin{aligned}&\mathrm{Sat}(C)\setminus\bigcup\{H : H\in\mathcal{H}_c(C),\, |H|>\varepsilon\} = \\ &\mathrm{Sat}(C)\setminus(\bigcup\{H : H\in\mathcal{H}_c(C),\, |H|>\varepsilon\}\cup H_\infty)\subseteq \\ &\mathrm{Sat}(C')\setminus\bigcup\{H' : H'\in\mathcal{H}_c(C'),\, |H'|>\varepsilon\}\end{aligned}$$

We conclude that $G_\varepsilon C\subseteq G_\varepsilon C'$ and, therefore, $G_\varepsilon X\subseteq G_\varepsilon Y$. □

Lemma 3.21. *Let $A, B\subseteq\overline{\Omega}$ be closed sets such that $A\cap B=\emptyset$. Then*

$$G_\varepsilon A\,\cap\, G_\varepsilon B=\emptyset. \tag{3.12}$$

Proof. Taking components of A and B instead of A and B, we may assume that A and B are connected. The result being obvious if $G_\varepsilon A$, or $G_\varepsilon B$, is empty, we may also assume that none of them is empty (thus, in particular, $|\mathrm{Sat}(A)|\geq\varepsilon$, $|\mathrm{Sat}(B)|\geq\varepsilon$). Observe that, by Lemma 2.11.(i), $\mathrm{Sat}(A)$ and $\mathrm{Sat}(B)$ are either nested or disjoint. If they are disjoint, since $G_\varepsilon A\subseteq\mathrm{Sat}(A)$ and $G_\varepsilon B\subseteq\mathrm{Sat}(B)$, (3.12) is obvious. If they are nested, without loss of generality, we may assume that $\mathrm{Sat}(A)\subseteq\mathrm{Sat}(B)$. Then $\mathrm{Sat}(A)$ is inside a hole H of B. We get that $|H|\geq|\mathrm{Sat}\,A|\geq\varepsilon$. Since $\mathrm{Sat}(A)$ is closed and H is open, $H\setminus\mathrm{Sat}(A)$ is an open and nonempty set, thus, it has positive measure. This yields $|H|>\varepsilon$, and therefore $H\cap G_\varepsilon B=\emptyset$. Hence $\mathrm{Sat}(A)\cap G_\varepsilon B=\emptyset$, which implies (3.12). □

Lemma 3.22. *Let X be a closed subset of $\overline{\Omega}$. Then*

$$Z\in\mathcal{CC}(G_\varepsilon X)\Leftrightarrow Z\neq\emptyset \text{ and } \exists Y\in\mathcal{CC}(X) \text{ such that } Z=G_\varepsilon Y.$$

Proof. Let $\mathcal{CC}(X)=\{C_i\}_{i\in I}$ and let $I_\varepsilon\subseteq I$ be the set of indices for which $|\mathrm{Sat}(C_i)|\geq\varepsilon$.

As we already observed we have the identity $G_\varepsilon X=\bigcup_{i\in I_\varepsilon}G_\varepsilon C_i$. We also observe that, since C_i is closed, each hole H of C_i is open and closed in $\overline{\Omega}\setminus C_i$, and therefore $C_i\cup H$ is connected ([84], IV, Thm. 3.4). Now, $G_\varepsilon C_i$, if not $\emptyset$, is the union of C_i and some of its internal holes and, thus, it is connected.

Let J be a subset of I_ε whose cardinal is at least 2, and let $D=\bigcup_{j\in J}G_\varepsilon C_j$. We now prove that D is not connected.

Since $D'=\bigcup_{j\in J}C_j$ is not connected, by the definition of connectedness, there is an open and closed subset E' in D' different from $\emptyset$ and D'. Observe that E' is an union of C_j, since each of those sets is connected. To be precise, let us write $E'=\bigcup_{l\in L}C_l$, with $\emptyset\neq L\subsetneq J$. Let $E=\bigcup_{l\in L}G_\varepsilon C_l$. Note that

$$E=E'\cup_{l\in L}\{\cup H : H\in\mathcal{H}_c(C_l),\, |H|\leq\varepsilon\}. \tag{3.13}$$

We prove that E is open and closed in D, hence, D is not connected.

Clearly, we have that $\emptyset \neq E \subsetneq D$. Since E' is open in D', there is an open set U' such that $U' \cap D' = E'$. Let U'' be the union of the connected components of U' that meet (thus, contain) some C_l, $l \in L$. Since $E' \subseteq U'$, we have that $E' \subseteq U''$. Let $U = U'' \cup E$. Observe that

$$\begin{aligned} U &= U'' \cup E = U'' \cup E' \cup_{l \in L} \{\cup H : H \in \mathcal{H}_c(C_l), \, |H| \leq \varepsilon\} \\ &= U'' \cup_{l \in L} \{\cup H : H \in \mathcal{H}_c(C_l), \, |H| \leq \varepsilon\} \end{aligned}$$

The set U'', as a union of open sets, is open, and the set U, as the union of U'' and some internal holes of C_l, which are all open, is also open. If $j \in J \setminus L$, by Lemma 3.21 we have that $E \cap G_\varepsilon C_j = \emptyset$. Let us prove that also $U'' \cap G_\varepsilon C_j = \emptyset$. Otherwise we would have

$$U'' \cap G_\varepsilon C_j \neq \emptyset, \tag{3.14}$$

and, by definition of U'' there would be a connected component O of U' containing C_l, for some $l \in L$, and $G_\varepsilon C_j$. We note that also $O \cap C_j \neq \emptyset$. Indeed, if $O \cap C_j = \emptyset$, then there is some internal hole H of C_j with $|H| \leq \varepsilon$ such that $O \cap H \neq \emptyset$. Let us prove that H cannot contain C_l. If $C_l \subseteq H$, then $\mathrm{Sat}(C_l) \subseteq H$ and, since $H \setminus \mathrm{Sat}(C_l)$ is open and not empty, thus of positive measure, we would have $|H| > |\mathrm{Sat}(C_l)| \geq \varepsilon$, a contradiction. Thus C_l is not contained in H. Hence the hole H_l of C_j containing C_l does not coincide with H. Since O is connected, $O \cap H \neq \emptyset$ and $O \cap H_l \neq \emptyset$, then O must intersect C_j. Thus we cannot assume that $O \cap C_j = \emptyset$ and, indeed, O meets C_j. This implies that $U' \cap C_j \neq \emptyset$ with $j \in J \setminus L$, contradicting the identity $U' \cap D' = E'$, which means that U' only intersects the C_i with $i \in L$. This contradiction proves that (3.14) does not hold, i.e., we have that $U'' \cap G_\varepsilon C_j = \emptyset$ for any $j \in J \setminus L$. Therefore $U'' \cap (D \setminus E) = U'' \cap \cup_{l \in J \setminus L} C_l = \emptyset$, and as a consequence

$$U \cap D = (U'' \cup E) \cap D = (U'' \cap D) \cup (E \cap D) = (U'' \cap E) \cup E = E.$$

We have shown that the set E is open in D. Applying the same argument to $D \setminus E$ instead of D we prove that E is also closed in D. We have shown that D is not connected. Thus the components of $G_\varepsilon X$ are the sets $G_\varepsilon C_i$. □

Lemma 3.23. *Let $(C_n)_{n \in \mathbb{N}}$ be a nonincreasing sequence of continua and $C = \bigcap_n C_n$. If H is a hole of C, there exists $n_0 \in \mathbb{N}$ and a nondecreasing sequence $(H_n)_{n \geq n_0}$, each H_n being a hole of C_n, such that $H = \bigcup_{n \geq n_0} H_n$.*

Proof. If $x \in H$, then there is some n_0 such that $x \notin C_n$ for $n \geq n_0$. Thus, if $n \geq n_0$, x is in some hole H_n of C_n. Obviously, $H_n \subseteq H$, and therefore $\bigcup_{n \geq n_0} H_n \subseteq H$.

Let $y \in \partial \bigcup_{n \geq n_0} H_n$, U be a neighborhood of y and V be a connected neighborhood of y such that $\overline{V} \subseteq U$. Since $y \in \overline{\bigcup_{n \geq n_0} H_n}$ there is some $n_1 \geq n_0$ such that

$$H_{n_1} \cap V \neq \emptyset;$$

and, by monotonicity of (H_n), we may write

$$H_n \cap V \neq \emptyset, \quad \forall n \geq n_1. \tag{3.15}$$

Since $y \in \overline{\overline{\Omega} \setminus \cup_{n \geq n_0} H_n} \subseteq \bigcap_{n \geq n_1} \overline{\overline{\Omega} \setminus H_n}$, we also have that

$$V \cap (\overline{\Omega} \setminus H_n) \neq \emptyset, \quad \forall n \geq n_1. \tag{3.16}$$

From (3.15) and (3.16), the connectedness of V implies that

$$V \cap \partial H_n \neq \emptyset, \quad \forall n \geq n_1.$$

Since $\partial H_n \subseteq C_n$, we get that $V \cap C_n \neq \emptyset$ for all $n \geq n_1$, and, thus, $\overline{V} \cap C_n \neq \emptyset$, for all $n \geq n_1$. It follows that $\overline{V} \cap C \neq \emptyset$, which implies that $U \cap C \neq \emptyset$. This being true for any neighborhood U of y, we get that $y \in \overline{C} = C$. Thus

$$\partial \bigcup_{n \geq n_0} H_n \subseteq C.$$

This implies that $H \cap \partial \bigcup_{n \geq n_0} H_n = \emptyset$, and, thus, $\bigcup_{n \geq n_0} H_n$ is closed in H. Since it is also open, as a union of open sets, the connectedness of H yields $H = \bigcup_{n \geq n_0} H_n$. □

Theorem 3.24. *The operator G_ε is upper semicontinuous on closed sets, i.e., if $(K_n)_{n \in \mathbb{N}}$ is a nonincreasing sequence of closed subsets of $\overline{\Omega}$, then*

$$G_\varepsilon \bigcap_{n \in \mathbb{N}} K_n = \bigcap_{n \in \mathbb{N}} G_\varepsilon K_n.$$

Proof. The inclusion of the left hand side term into the right hand side one is due to the monotonicity of G_ε. We just have to show the other inclusion.

Let $x \in \bigcap_n G_\varepsilon K_n$. Using Lemma 3.22, for any n, there is some $C_n \in \mathcal{CC}(K_n)$ such that $x \in G_\varepsilon C_n$. For each $n \geq 1$, C_n is inside some $Q \in \mathcal{CC}(K_{n-1})$, and then $G_\varepsilon C_n \subseteq G_\varepsilon Q$. Thus $x \in G_\varepsilon(Q) \cap G_\varepsilon(C_{n-1}) \neq \emptyset$, implying, by Lemma 3.22, that $Q = C_{n-1}$. This shows that the sequence of continua (C_n) is nonincreasing, so that their intersection C is a continuum ([84], Thm. 5.3). Since $C \subseteq \bigcap_n K_n$, the desired result will be implied if we prove that $x \in G_\varepsilon C$.

By Lemma 2.13, we have that $\operatorname{Sat} C = \bigcap_n \operatorname{Sat} C_n$. Thus $|\operatorname{Sat} C| \geq \varepsilon$, hence $C \subseteq G_\varepsilon C$. Since $x \in G_\varepsilon C_n \subseteq \operatorname{Sat}(C_n)$, we have that $x \in \operatorname{Sat}(C)$. If $x \in C$, then $x \in G_\varepsilon C$. If $x \notin C$, there is an internal hole of C, say H, such that $x \in H$. By Lemma 3.23, there is a nondecreasing sequence H_n of internal holes of C_n such that $H = \cup_n H_n$ and we may assume that $x \in H_n$. Since $x \in G_\varepsilon C_n$, the sequence H_n is such that $|H_n| \leq \varepsilon$ and thus $|H| = |\bigcup H_n| \leq \varepsilon$. Hence $x \in H \subseteq G_\varepsilon C$. □

3.4.3 Properties

Lemma 3.25. *Let A be a closed subset of Ω such that $p_\infty \notin A$. Assume that $x \in \overline{\Omega} \setminus G_\varepsilon A$. Then $D = \bigcap_{C \in \mathcal{CC}(A)} \overline{\overline{\Omega} \setminus \mathrm{Sat}(C)}$ is a continuum contained in $\overline{\overline{\Omega} \setminus A}$ such that $p_\infty \in D$.*

Proof. For any $C \in \mathcal{CC}(A)$, since $p_\infty \notin C$, $\overline{\Omega} \setminus \mathrm{Sat}\, C$ is a nonempty and connected set, and, thus, the set $\overline{\overline{\Omega} \setminus \mathrm{Sat}\, C}$ is a continuum. By Lindelöf's theorem, the intersection of a family of continua can be written as the intersection of a sequence of them. We can thus find a sequence (C_n) of connected components of A such that

$$\bigcap_{C=\mathrm{cc}(A)} \overline{\overline{\Omega} \setminus \mathrm{Sat}(C)} = \bigcap_{n \in \mathbb{N}} \overline{\overline{\Omega} \setminus \mathrm{Sat}(C_n)}.$$

Let $D_n = \bigcap_{k=0}^{n} \overline{\overline{\Omega} \setminus \mathrm{Sat}\, C_k}$. Clearly, (D_n) is a nonincreasing sequence of continua, and we may write

$$\bigcap_{C=\mathrm{cc}(A)} \overline{\overline{\Omega} \setminus \mathrm{Sat}(C)} = \bigcap_{n \in \mathbb{N}} D_n,$$

which is thus a continuum D. Obviously $p_\infty \in D$. Let $F = \overline{\overline{\Omega} \setminus A}$. Let us prove that $D \subseteq F$. If this is not the case, there is $y \in D \setminus F \subseteq A$, and there exists $C \in \mathcal{CC}(A)$ with $y \in C$. Since $y \in D \subseteq \overline{\overline{\Omega} \setminus \mathrm{Sat}(C)}$, y is not in the interior of $\mathrm{Sat}(C)$ while it is in C. Thus $y \in \partial\, \mathrm{Sat}(C) \subseteq \partial C$. Now, we prove that (*) for any connected neighborhood U of y, $U \cap F \neq \emptyset$. This will imply that $y \in \overline{F} = F$, a contradiction. We conclude that $D \subseteq F$. To prove (*), assume on the contrary that U is a connected neighborhood of y such that $U \cap F = \emptyset$. Then $U \subseteq A$, and also $C \cup U \subseteq A$. Since $y \in C \cap U$, the set $C \cup U$ is connected and this implies that $U \subseteq C$. On the other hand, since $y \in \partial C$, $U \cap (\overline{\Omega} \setminus C) \neq \emptyset$. This contradiction proves (*) and the Lemma follows. □

Lemma 3.26. *Let $A, B \subseteq \overline{\Omega}$ be closed sets such that $A \cup B = \overline{\Omega}$. Then*

$$G_\varepsilon A \,\cup\, G_\varepsilon B = \overline{\Omega}. \tag{3.17}$$

Proof. Let $x \notin G_\varepsilon A$. We shall prove that $x \in G_\varepsilon B$. Since $x \notin G_\varepsilon A$, one of the following cases happens:

(i) $\forall C \in \mathcal{CC}(A)$, $x \notin \mathrm{Sat}(C)$.
(ii) $\forall C \in \mathcal{CC}(A)$, $x \in \mathrm{Sat}(C) \Rightarrow |\,\mathrm{Sat}(C)| < \varepsilon$.
(iii) There is some $C \in \mathcal{CC}(A)$ such that $x \in \mathrm{Sat}(C)$ and $|\,\mathrm{Sat}(C)| \geq \varepsilon$.

Suppose that (i) holds. Observe that in this case $p_\infty \notin A$, since, otherwise, there would be a connected component C_∞ of A containing p_∞. As a consequence $x \in \overline{\Omega} = \mathrm{Sat}(C_\infty)$. Let D be the set defined in the Lemma 3.25. Then $x \in D$ and $D \subseteq B$. Since $\mathrm{Sat}(D) = \overline{\Omega}$, we have $|\mathrm{Sat}(D)| \geq \varepsilon$, and, therefore, $x \in D \subseteq G_\varepsilon D \subseteq G_\varepsilon B$.

Suppose that *(ii)* holds. As above we observe that $p_\infty \notin A$. Let D be the set defined in Lemma 3.25. Let D_B be the connected component of B containing D. If $x \in D_B$ we have again that $x \in D_B \subseteq G_\varepsilon D_B \subseteq G_\varepsilon B$, and we are done. If $x \notin D_B$, since $\mathrm{Sat}(D_B) = \overline{\Omega}$, then x is in a hole H of D. Thanks to Proposition 3.14, we may write

$$H = \bigcup_{G=\mathrm{Sat}(G'), G' \in \mathcal{CC}(\overline{\Omega} \setminus B), x \in G \subseteq H} G.$$

Thanks to Lindelöf's theorem, we may write $H = \cup_n G_n$ where, for each n, $G_n = \mathrm{Sat}(G'_n)$, $G'_n \in \mathcal{CC}(\overline{\Omega} \setminus B)$, and $x \in G_n \subseteq H$. Since $\overline{\Omega} \setminus B \subseteq A$, we have that $G'_n \subseteq A$ and there is a connected component C of A such that $G'_n \subseteq C$. Therefore $G_n = \mathrm{Sat}(G'_n) \subseteq \mathrm{Sat}(C)$ and we obtain that $|G_n| \leq |\mathrm{Sat}(C)| < \varepsilon$. We conclude that $|H| = \sup_{n \in \mathbb{N}} |G_n| \leq \varepsilon$. Thus $x \in H \subseteq G_\varepsilon D_B \subseteq G_\varepsilon B$.

Finally, we suppose that *(iii)* holds. Let $\mathcal{F} = \{E = \mathrm{Sat}(C) : C \in \mathcal{CC}(A), x \in E, |E| \geq \varepsilon\}$ and

$$T = \bigcap_{E \in \mathcal{F}} C.$$

We claim that $T = \mathrm{Sat}(T')$, with $T' \in \mathcal{CC}(A)$. If $\mathcal{F}$ contains a finite number of sets, our claim is true since they are nested. Thus, we may assume that $\mathcal{F}$ contains an infinite number of sets. Thanks to Lindelöf's theorem, we may write their intersection as the intersection of a sequence E_n of them. By taking $\bigcap_{k=0}^n E_k \in \mathcal{F}$ instead of E_n, we may also assume that this sequence is nonincreasing. The set T is therefore a continuum and $|T| = \inf_{n \in \mathbb{N}} |E_n| \geq \varepsilon$. Let us prove that $\partial T \subseteq A$. For that, let $y \in \partial T$ and let U a connected neighborhood of y. Then $U \cap T \neq \emptyset$ and $\cup_n(\overline{\Omega} \setminus E_n) = U \cap (\overline{\Omega} \setminus T) \neq \emptyset$. There is some n_0 such that $U \cap (\overline{\Omega} \setminus E_{n_0}) \neq \emptyset$. Since also $U \cap E_{n_0} \neq \emptyset$, we have that $U \cap \partial E_{n_0} \neq \emptyset$. Since $\partial E_{n_0} \subseteq \partial A \subseteq A$, we conclude that $U \cap A \neq \emptyset$, and, therefore $y \in \overline{A} = A$. This proves that $\partial T \subseteq A$. On the other hand, we observe that the complement of T is connected, being the union of an increasing sequence of connected sets. Since $\overline{\Omega}$ is unicoherent, and $\partial T = T \cap (\overline{\overline{\Omega} \setminus T})$, being the intersection of two continua, it is connected. Thus, there is some $T' \in \mathcal{CC}(A)$ such that $\partial T \subseteq T'$. If $p_\infty \in T$, then $p_\infty \in E$ for any $E \in \mathcal{F}$. Then $E = \overline{\Omega}$ for any $E \in \mathcal{F}$ and, thus, $T = \overline{\Omega} = \mathrm{Sat}(C)$ for any $\mathrm{Sat}(C) \in \mathcal{F}$. If $p_\infty \notin T$, then $T \neq \overline{\Omega}$ and we may use Lemma 2.10 to obtain that $T = \mathrm{Sat}(\partial T) \subseteq \mathrm{Sat}(T')$. Since $\partial T \subseteq T'$ and $\partial T \subseteq T \subseteq E$ for any $E \in \mathcal{F}$, we have that $T' \cap E \neq \emptyset$, and therefore $\mathrm{Sat}(T') \subseteq E$ for any $E \in \mathcal{F}$. It follows that $\mathrm{Sat}(T') \subseteq T$. We have, thus, the equality $T = \mathrm{Sat}(T')$ and $T \in \mathcal{F}$. This implies that $T' \subseteq G_\varepsilon T' \subseteq G_\varepsilon A$, and, thus, $x \notin T'$.

Since $x \in T$, then x must be in a hole H of T' and we must have $|H| > \varepsilon$, since otherwise $x \in H \subseteq G_\varepsilon T' \subseteq G_\varepsilon A$. Using Proposition 3.14, we write

$$H = \bigcup_{G=\mathrm{Sat}(G'), G' \in \mathcal{CC}(\overline{\Omega}\setminus A),\, x \in G \subseteq H} G,$$

and, again, using Lindelöf's theorem, we may write the above union as the union of a sequence G_n, which we may take as nondecreasing since the sets G_n are saturated and all contain x. Since $|H| > \varepsilon$, there is some $n_0 \in \mathbb{N}$ such that $|G_n| > \varepsilon$ for all $n \geq n_0$. Fix such a value of n. We write $G_n = \mathrm{Sat}(G'_n)$, where $G'_n \in \mathcal{CC}(\overline{\Omega}\setminus A)$ and $x \in G_n \subseteq H$. Let $Q \in \mathcal{CC}(B)$ such that $G'_n \subseteq Q$. If $x \in G'_n$, then also $x \in G_\varepsilon G'_n \subseteq G_\varepsilon Q \subseteq G_\varepsilon B$. If $x \notin G'_n$, then x is in a hole H' of G'_n, and, since $\partial H' \subseteq \partial G'_n \subseteq \partial A \subseteq A$, there is $K \in \mathcal{CC}(A)$ containing $\partial H'$. Observe that K intersects H', a hole of G'_n. Since $p_\infty \notin H'$, by Lemma 2.10, we have that $H' = \mathrm{Sat}(\partial H') \subseteq \mathrm{Sat}(K)$. Since $K \subseteq A$ and $G' \subseteq \overline{\Omega}\setminus A$, K and G'_n cannot intersect. Hence K cannot intersect another hole of G'_n, since it would intersect G'_n. Thus $K \subseteq H'$, and, also $\mathrm{Sat}(K) \subseteq H' \subsetneq \mathrm{Sat}(G'_n) \subseteq T$. In particular, $H' = \mathrm{Sat}(K) \notin \mathcal{F}$, and this implies that $|H'| < \varepsilon$. We conclude that $H' \subseteq G_\varepsilon G'_n$ and, thus, $x \in G_\varepsilon G'_n$. Since G'_n is contained in a connected component of B also $x \in G_\varepsilon B$. □

Theorem 3.27. *The operator G_ε may be lifted to an operator $\tilde{G}_\varepsilon$ acting on upper semicontinuous functions. Moreover $\tilde{G}_\varepsilon$ maps upper semicontinuous functions into upper semicontinuous functions.*

Proof. By Lemma 3.24, and the results recalled in Sect. 3.2.2, G_ε may be lifted to an operator defined on upper semicontinuous functions by the formula $[\tilde{G}_\varepsilon u \geq \lambda] = G_\varepsilon[u \geq \lambda]$, $\lambda \in \mathbb{R}$. To prove that $\tilde{G}_\varepsilon u$ is upper semicontinuous, it suffices to prove that $G_\varepsilon[u \geq \lambda]$ is a closed set. Thus, it will be sufficient to prove that the image by G_ε of a closed set F is also closed. Observe that this is true if F is closed and connected. Indeed, in this case, the complement of $G_\varepsilon F$ is a union of holes of F which are open, and, thus, $G_\varepsilon F$ is closed. Assume now that F is a general closed subset of $\overline{\Omega}$. Let $(x_n)_{n\in\mathbb{N}}$ a sequence of points of $G_\varepsilon F$ converging to $x \in \overline{\Omega}$. We shall prove that $x \in G_\varepsilon F$.

As shown by Lemma 3.22, x_n belongs to some $G_\varepsilon F_n$, where $F_n \in \mathcal{CC}(F)$. Obviously, we may assume that $x \notin G_\varepsilon F_n$, for all n. If the family $\{F_n,\, n \in \mathbb{N}\}$ is finite, we can extract a subsequence of $(x_n)_{n\in\mathbb{N}}$ belonging to some $G_\varepsilon F_{n_0}$ which is closed. Thus, $x \in G_\varepsilon F_{n_0} \subseteq G_\varepsilon F$. We may now assume that $\{F_n,\, n \in \mathbb{N}\}$ is infinite, and, possibly after extraction of a subsequence, we may also assume that $F_m \cap F_n = \emptyset$ for any $m \neq n$.

Since $\mathrm{Sat}(G_\varepsilon F_n) = \mathrm{Sat}(F_n)$, we have that $|\mathrm{Sat}(F_n)| \geq \varepsilon$. This implies that only a finite number of these saturations are two by two disjoint. Thus, after extraction of a subsequence, if necessary, we may assume that they all intersect, so that they form either a decreasing or an increasing sequence.

If the sequence $(\mathrm{Sat}(F_n))_{n\in\mathbb{N}}$ is decreasing, then their intersection is a set $\mathrm{Sat}(F')$, $F' \in \mathcal{CC}(K)$. This can be shown as in Lemma 3.26. Then we have that $|\mathrm{Sat}(F')| = \inf_n |\mathrm{Sat}(F_n)| \geq \varepsilon$. Since $F' \cap F_n = \emptyset$ for all n (otherwise, we would have $F' = F_n$ for all sufficiently large n), and $F' \subseteq \mathrm{Sat}(F_n)$, we have that $\mathrm{Sat}(F')$ is contained in an internal hole H_n of F_n for any n. Since

H_n is open, we have that $|H_n| > \varepsilon$ and then $H_n \cap G_\varepsilon F_n = \emptyset$. This implies that $x_n \notin \mathrm{Sat}(F')$ for any n. Since x_n converges to x and $x \in \mathrm{Sat}(F')$ we conclude that $x \in \partial\,\mathrm{Sat}(F') \subseteq \partial F' \subseteq F' \subseteq G_\varepsilon F' \subseteq G_\varepsilon F$, the desired result.

Let us assume now that $(\mathrm{Sat}(F_n))_{n\in\mathbb{N}}$ is increasing. We claim that $x \in \liminf F_n$. This is obviously true if $x_n \in F_n$. If $x_n \notin F_n$, then there is an internal hole H_n of F_n such that $|H_n| \leq \varepsilon$ and $x \in F_n$. Let V be a connected neighborhood of x. For n large enough, $x_n \in V$, thus $H_n \cap V \neq \emptyset$. Since $x \notin G_\varepsilon F_n$, V also intersects another hole of F_n. Hence $V \cap F_n \neq \emptyset$. It follows that $x \in \liminf F_n$. The lim inf of F_n being nonempty, its lim sup is a continuum C [51], 42.II.6. Since F is compact, it follows that $C \subseteq F$. Let $F' = \mathrm{cc}(F, C)$. We observe that $x \in F'$. If $F_n \cap F' \neq \emptyset$, then $F_n = F'$ since both are connected components of F. Since we are assuming that the sets F_n are two by two disjoint, this cannot happen more than once. Thus we may assume that $F_n \cap F' = \emptyset$ for all n. The sequence $(\mathrm{Sat}(F_n))$ being increasing, all F_n are in the same hole H of F'. This hole must be internal. Suppose that H is the external hole of F'. Since H is open, there is a continuum L contained in H joining p_∞ and an arbitrary point y_0 of F_0. Since F_0 is in an internal hole of F_1, there is some $y_1 \in L \cap F_1$. In this manner, we can construct a sequence $(y_n)_{n\in N}$ such that $y_n \in L \cap F_n$ for all n. L being compact, some subsequence of (y_n) converges to a point $y \in L \cap C$. It follows that $F' \cap L \neq \emptyset$, contrary to the assumption that L was contained in H. This proves that H is internal. Since $G_\varepsilon F_n \subseteq \mathrm{Sat}(F_n) \subseteq H \subseteq \mathrm{Sat}(F')$, we have that $|\mathrm{Sat}(F')| \geq \varepsilon$, thus $x \in F' \subseteq G_\varepsilon F' \subseteq G_\varepsilon F$. □

Theorem 3.28. *$\tilde{G}_\varepsilon$ maps continuous functions to continuous functions and it is self-dual when acting on them.*

Proof. First we shall prove that $\tilde{G}_\varepsilon(-u) = -\tilde{G}_\varepsilon(u)$ for any function $u \in C(\overline{\Omega})$. For that we shall prove that condition 3 of Definition 3.19 holds.

According to (3.12), we may write for any $\lambda \in \mathbb{R}$,

$$\forall n > 0, \quad G_\varepsilon[u \leq \lambda] \cap G_\varepsilon[u \geq \lambda + \frac{1}{n}] = \emptyset,$$

and, since $G_\varepsilon[u \geq \lambda + \frac{1}{n}] = [\tilde{G}_\varepsilon u \geq \lambda + \frac{1}{n}]$, by taking the intersection over all n, we get

$$G_\varepsilon[u \leq \lambda] \cap [\tilde{G}_\varepsilon u > \lambda] = \emptyset.$$

Therefore

$$G_\varepsilon[u \leq \lambda] \subseteq [\tilde{G}_\varepsilon u \leq \lambda].$$

Now, let x be such that for all $n > 0$, $x \notin G_\varepsilon[u \geq \lambda + \frac{1}{n}]$. Due to (3.17), we have that $x \in \bigcap_n G_\varepsilon[u \leq \lambda + \frac{1}{n}]$. Since G_ε is upper semicontinuous,

$$x \in G_\varepsilon \bigcap_n [u \leq \lambda + \frac{1}{n}] = G_\varepsilon[u \leq \lambda].$$

This proves that

$$G_\varepsilon[u \leq \lambda] \cup \bigcup_n G_\varepsilon[u \geq \lambda + \frac{1}{n}] = \overline{\Omega},$$

and, thus,

$$G_\varepsilon[u \leq \lambda] \cup \bigcup_n [\tilde{G}_\varepsilon u \geq \lambda + \frac{1}{n}] = \overline{\Omega}.$$

By taking the complement of each part, we get:

$$(\overline{\Omega} \setminus G_\varepsilon[u \leq \lambda]) \cap [\tilde{G}_\varepsilon u \leq \lambda] = \emptyset,$$

which proves that

$$[\tilde{G}_\varepsilon u \leq \lambda] \subseteq G_\varepsilon[u \leq \lambda],$$

and, actually, we have the equality of both sets.

Applying Theorem 3.27, if u is continuous, $\tilde{G}_\varepsilon u$ is upper semicontinuous. Since $-u$ is also upper semicontinuous, $\tilde{G}_\varepsilon(-u) = -\tilde{G}_\varepsilon u$ is also upper semicontinuous, hence $\tilde{G}_\varepsilon u$ is lower semicontinuous. Thus, $\tilde{G}_\varepsilon u$ is continuous. □

Proposition 3.29. *For $\varepsilon' \geq \varepsilon$, $\tilde{G}_{\varepsilon'} = \tilde{G}_{\varepsilon'} \circ \tilde{G}_\varepsilon$. Therefore $\tilde{G}_\varepsilon$ is idempotent.*

Proof. The conclusion that $\tilde{G}_\varepsilon$ is idempotent derives from the previous statement by taking $\varepsilon' = \varepsilon$.

The result amounts to show that for any λ, $G_{\varepsilon'}[u \geq \lambda] = G_{\varepsilon'} G_\varepsilon[u \geq \lambda]$. We distinguish three families among the connected components $(C_i)_{i\in I}$ of $[u \geq \lambda]$:

$$\begin{aligned} I_{\varepsilon'} &= \{i \in I : \, |\operatorname{Sat}(C_i)| \geq \varepsilon'\}, \\ I_\varepsilon &= \{i \in I : \, \varepsilon' > |\operatorname{Sat}(C_i)| \geq \varepsilon\}, \\ I_0 &= \{i \in I : \, \varepsilon > |\operatorname{Sat}(C_i)|\}. \end{aligned}$$

We observe that $I_\varepsilon = \emptyset$ if $\varepsilon = \varepsilon'$. Thanks to Lemma 3.22, we may write

$$\begin{aligned} G_\varepsilon[u \geq \lambda] = & \bigcup_{i \in I_{\varepsilon'}} \{\operatorname{Sat}(C_i) \setminus \bigcup H, \, H \text{ internal hole of } C_i, \, |H| > \varepsilon\} \cup \\ & \bigcup_{i \in I_\varepsilon} \{\operatorname{Sat}(C_i) \setminus \bigcup H, \, H \text{ internal hole of } C_i, \, |H| > \varepsilon\}. \end{aligned}$$

If $i \in I_\varepsilon$, the measure of

$$\operatorname{Sat}(\operatorname{Sat}(C_i) \setminus \bigcup H) = \operatorname{Sat}(\operatorname{Sat}(C_i)) = \operatorname{Sat}(C_i)$$

is $< \varepsilon'$ and

$$G_{\varepsilon'}(\operatorname{Sat}(C_i) \setminus \bigcup H) = \emptyset.$$

For the same reason, if $i \in I_{\varepsilon'}$, we have $|\operatorname{Sat}(\operatorname{Sat}(C_i) \setminus \bigcup H)| \geq \varepsilon'$. Thus,

$$G_{\varepsilon'}(\operatorname{Sat}(C_i) \setminus \bigcup_{|H|>\varepsilon} H) = \operatorname{Sat}(C_i) \setminus \bigcup_{|H|>\varepsilon'} H = G_{\varepsilon'} C_i.$$

In conclusion, this yields

$$G_{\varepsilon'} G_{\varepsilon} C_i = G_{\varepsilon'} C_i. \qquad \square$$

The following properties are an easy consequence of the definition of $\tilde{G}_\varepsilon$.

Lemma 3.30. *The operator $\tilde{G}_\varepsilon$ satisfies the following properties*

(i) If $u \leq v$, then $\tilde{G}_\varepsilon u \leq \tilde{G}_\varepsilon v$, for any $u, v \in \mathcal{USC}(\overline{\Omega})$.
(ii) $\tilde{G}_\varepsilon(\alpha) = \alpha$ for any $\alpha \in \mathbb{R}$.
(iii) $\tilde{G}_\varepsilon(u + \alpha) = \tilde{G}_\varepsilon(u) + \alpha$ for any function $u \in \mathcal{USC}(\overline{\Omega})$ and any $\alpha \in \mathbb{R}$.
(iv) $\|\tilde{G}_\varepsilon(u) - \tilde{G}_\varepsilon(v)\|_\infty \leq \|u - v\|_\infty$ for any $u, v \in \mathcal{USC}(\overline{\Omega})$.

Thus, $\tilde{G}_\varepsilon$ is increasing and idempotent, it is indeed a filter.

Lemma 3.31. *Let u be an upper semicontinuous function and let $v = M_\delta^+ M_\delta^- u$. Then $\tilde{G}_\varepsilon v = v$ for any $\varepsilon < \delta$.*

Proof. Let us fix $\varepsilon < \delta$. Since $[\tilde{G}_\varepsilon v \geq \lambda] = G_\varepsilon[v \geq \lambda]$, it will be sufficient to prove that $G_\varepsilon[v \geq \lambda] = [v \geq \lambda]$ for almost all $\lambda \in \mathbb{R}$. Let $K = [v \geq \lambda]$. We observe that

$$G_\varepsilon K = \bigcup_{C \in \mathcal{CC}(K), |\operatorname{Sat}(C)| \geq \varepsilon} \{\operatorname{Sat}(C) \setminus \bigcup H,\ H \text{ internal hole of } C,\ |H| > \varepsilon\}.$$

Since, by Proposition 3.11, the connected components of K have measure $\geq \delta$, any $C \in \mathcal{CC}(K)$ satisfies $|\operatorname{Sat}(C)| \geq \varepsilon$. Similarly, since any internal hole of any $C \in \mathcal{CC}(K)$ contains a connected component of $[v < \lambda]$ ([10], Prop. 14), it has also measure $\geq \delta > \varepsilon$. Hence, the family of internal holes H of the connected components of K such that $|H| \leq \varepsilon$ is empty. We conclude that $G_\varepsilon K = K$ and, therefore, $G_\varepsilon v = v$. $\square$

Proposition 3.32. *Let $u \in C(\overline{\Omega})$. Then $\tilde{G}_\varepsilon u \to u$ uniformly as $\varepsilon \to 0+$.*

Proof. Using Lemma 3.31, we have

$$\tilde{G}_\varepsilon u - u = \tilde{G}_\varepsilon u - \tilde{G}_\varepsilon(M_\delta^+ M_\delta^- u) + M_\delta^+ M_\delta^- u - u.$$

for any $\varepsilon, \delta > 0$ such that $\varepsilon < \delta$. By Proposition 3.10, given $\alpha > 0$, there is some $\delta_0 > 0$ such that

$$u - \alpha \leq M_{\delta_0}^+ M_{\delta_0}^- u \leq u + \alpha.$$

This implies that

$$\tilde{G}_\varepsilon u - \alpha \leq \tilde{G}_\varepsilon(M^+_{\delta_0} M^-_{\delta_0} u) \leq \tilde{G}_\varepsilon u + \alpha \quad \forall \varepsilon > 0,$$

and, therefore,

$$|\tilde{G}_\varepsilon(M^+_{\delta_0} M^-_{\delta_0} u) - \tilde{G}_\varepsilon u| \leq \alpha \quad \forall \varepsilon > 0.$$

Now, we choose $\varepsilon < \delta_0$ and we obtain that

$$|\tilde{G}_\varepsilon u - u| \leq |\tilde{G}_\varepsilon(M^+_{\delta_0} M^-_{\delta_0} u) - \tilde{G}_\varepsilon u| + |M^+_{\delta_0} M^-_{\delta_0} u - u| \leq 2\alpha.$$

The proposition follows. □

3.4.4 The Effect of Grain Filter on the Tree of Shapes

Theorem 3.33. *Let $u : \overline{\Omega} \to \mathbb{R}$ be an upper semicontinuous function. We have*

(i) If X is a shape of upper type of $\tilde{G}_\varepsilon u$, then X is a shape of upper type of u.
(ii) If X is a shape of lower type of $\tilde{G}_\varepsilon u$, then X is a limit shape of lower type of u.

Proof. (*i*) Let X be a shape of upper type of $\tilde{G}_\varepsilon u$. Let $\lambda \in \mathbb{R}$, $Y \in \mathcal{CC}([\tilde{G}_\varepsilon u \geq \lambda])$ be such that $X = \mathrm{Sat}(Y)$. Since $[\tilde{G}_\varepsilon u \geq \lambda] = G_\varepsilon[u \geq \lambda]$, by Lemma 3.22, there is $Z \in \mathcal{CC}([u \geq \lambda])$ such that $Y = G_\varepsilon(Z)$. Since $Y \neq \emptyset$, we have

$$Y = Z \cup \cup\{H : H \text{ is a hole of } Z \text{ with } |H| \leq \varepsilon\}$$

and we conclude that $\mathrm{Sat}(Y) = \mathrm{Sat}(Z)$. Thus X is a shape of upper type of u.
(*ii*) Let X be a shape of lower type of $\tilde{G}_\varepsilon u$. Let $\lambda \in \mathbb{R}$, $Y \in \mathcal{CC}([\tilde{G}_\varepsilon u < \lambda])$ be such that $X = \mathrm{Sat}(Y)$. Again we observe that, if $X = \overline{\Omega}$, and u being upper bounded, we have that X is also a shape of lower type of u. Thus, we may assume that $X = \mathrm{Sat}(Y) \neq \overline{\Omega}$, and $p_\infty \notin Y$. Recall that

$$[\tilde{G}_\varepsilon u < \lambda] = \overline{\Omega} \setminus [\tilde{G}_\varepsilon u \geq \lambda] = \overline{\Omega} \setminus G_\varepsilon[u \geq \lambda].$$

Now, we use Lemma 3.15 with $A := G_\varepsilon[u \geq \lambda]$, and Y being the set we are considering here, to conclude that $\mathrm{Sat}(Y)$ is a hole of a connected component of A, i.e, there exists $Z \in \mathcal{CC}(G_\varepsilon[u \geq \lambda])$ such that $X = \mathrm{Sat}(Y)$ is a hole of Z. Now, we apply Lemma 3.22, and there is $Z' \in \mathcal{CC}([u \geq \lambda])$ such that $Z = G_\varepsilon(Z')$. Notice that X is also a hole of Z', i.e., X is a hole of a connected component of $[u \geq \lambda]$, call it C. As in Theorem 3.16.(ii), by Proposition 3.14, we conclude that X is a limit shape of lower type of u. □

The proof of the following result is analogous to the proof of Proposition 3.18 and we shall omit it.

Proposition 3.34. *Let $u : \overline{\Omega} \to \mathbb{R}$ be a bounded upper semicontinuous function. If X is a limit shape of $\tilde{G}_\varepsilon u$, then X a limit shape of u. In other words, the tree structure of $\tilde{G}_\varepsilon u$ is a simplified version of the tree structure of u.*

Remark 3.35. We observe that G_ε does not kill the oscillations. Indeed, if $u(x) = (\|x\| - 1)\sin \frac{1}{\|x\|-1}$ if $\|x\| \geq 1$, and $u(x) = 0$ if $\|x\| \leq 1$, where $x \in \mathbb{R}^N$ and $\|x\|$ denotes the euclidean norm of x, then $G_\varepsilon u = u$ for any $\varepsilon < \|B(0,1)\|$, $B(0,1)$ being the unit ball of $\mathbb{R}^N$.

3.5 Relations with Other Operators of Mathematical Morphology

3.5.1 Relations with Grain Operators

We want to compare the grain filter described above with the notion of grain operators as defined in [38, 45]. To fix ideas, we shall work in $\overline{\Omega}$ with the classical connectivity. Thus, we denote by $\mathcal{C}$ the family of connected sets of $\overline{\Omega}$. A *grain criterion* is a mapping $c : \mathcal{C} \to \{0,1\}$. Given two grain criteria, f for the foreground and b for the background, the associated *grain operator* $\psi_{f,b}$ is defined by

$$\psi_{f,b}(X) = \bigcup \{C : (C = cc(X) \text{ and } f(C) = 1) \\ \text{or } (C = cc(X^c) \text{ and } b(C) = 0)\}.$$

In [45], Heijmans characterizes grain operators that are self-dual and those that are increasing. Indeed, he proves that $\psi_{f,b}$ is self-dual if and only if $f = b$. He also proves that $\psi_{f,b}$ is increasing if and only if f and b are increasing and the following condition holds

$$\sup(f(cc(X \cup \{x\}, x)), b(cc(X^c \cup \{x\}, x)) = 1, \tag{3.18}$$

for any $X \subseteq \overline{\Omega}$ and any $x \in \overline{\Omega}$. There is an extensive literature on self-dual and on connected filters and we refer to [38, 44, 70, 71, 103] and the references therein.

Let us finish this section with a remark. We shall say that a grain criterion $c : \mathcal{C} \to \{0,1\}$ is upper semicontinuous on compact sets if $c(\cap_n K_n) = \inf_n c(K_n)$ for any decreasing sequence of continua K_n, $n \in \mathbb{N}$.

Proposition 3.36. *Let ψ be a self-dual and increasing grain operator in $\overline{\Omega}$ associated to the grain criterion c. Assume that c is upper semicontinuous. Then $\psi(X) = X$ for any $X \subseteq \overline{\Omega}$.*

Proof. We have that either $c(\{x\}) = 1$ for all $x \in \overline{\Omega}$, or $c(\{x\}) = 0$ for some point $x \in \overline{\Omega}$. In the first case, we deduce that $c(X) = 1$ for any nonempty subset X of $\overline{\Omega}$ and, therefore, we have that $\psi(X) = X$ for any $\emptyset \neq X \subseteq \overline{\Omega}$. Since also $\psi(\emptyset) = \emptyset$, we have that $\psi(X) = X$ for any $X \subseteq \overline{\Omega}$. In the second case, using the upper semicontinuity of c we have that $c(\overline{B(x,r)}) = 0$ for some $r > 0$ small enough. Then, we choose $X = \overline{B(x,r)}$ and observe that $c(cc(X \cup \{x\}, x)) = c(X) = 0$ and $c(cc(X^c \cup \{x\}, x)) = c(\{x\}) = 0$, which contradicts (3.18). □

The above proposition says that there are no nontrivial translation invariant, increasing and self-dual grain operators which are based on upper semicontinuous grain criteria. Other types of connected operators called flattenings and levelings were introduced by Meyer in [70,71] and further studied in [105]. In particular, Serra proves that there exist increasing and selfdual flattenings and levelings based on markers [105].

3.5.2 Relations with Connected Operators

Finally, let us prove that the grain filter we have introduced above corresponds to a universal criterion to define increasing and self-dual filters. Let us recall the definition of connected operator [45, 98, 106]. For that, given a set $X \subseteq \overline{\Omega}$, we denote by $P(X)$ the partition of $\overline{\Omega}$ constituted by the $cc(X)$ and $cc(X^c)$. The family of all subsets of X will be denoted by $\mathcal{P}(\overline{\Omega})$.

Definition 3.37. Let $\mathcal{F} \subseteq \mathcal{P}(\overline{\Omega})$. An operator $T : \mathcal{F} \to \mathcal{F}$ is connected if the partition $P(T(X))$ is coarser than $P(X)$ for every set $X \subseteq \overline{\Omega}$.

Let $\mathcal{B}(\overline{\Omega})$ be the family of Borel sets of $\overline{\Omega}$, i.e., the σ-algebra of subsets of $\overline{\Omega}$ generated by the open sets. Observe that, if $X \in \mathcal{B}(\overline{\Omega})$, then the connected components of X are also in $\mathcal{B}(\overline{\Omega})$. Given a connected operator $T : \mathcal{B}(\overline{\Omega}) \to \mathcal{B}(\overline{\Omega})$ we shall say that

(i) T is increasing if $T(X) \subseteq T(Y)$ for any $X, Y \in \mathcal{B}(\overline{\Omega})$ such that $X \subseteq Y \subseteq \overline{\Omega}$.
(ii) ψ acts additively on connected components if $T(X) = \cup_i T(X_i)$ when X_i is the family of connected components of $X \in \mathcal{B}(\overline{\Omega})$.
(iii) T is self-dual if $T(\overline{\Omega} \setminus X) = \overline{\Omega} \setminus T(X)$ for any open or closed set $X \subseteq \overline{\Omega}$.
(iv) T is bounded if $p_\infty \notin T(X)$ when $p_\infty \notin X$.

We have restricted the notion of self-duality of T to open and closed sets for convenience. Indeed this will be sufficient for our purposes, which are to extend the self-duality of the set operator to continuous functions. In the context of Mathematical Morphology this notion could be defined for any subset of $\overline{\Omega}$.

There are explicit characterizations of self-dual filters in the mathematical morphology literature [38, 44]. In our context, we can also give an explicit formula for connected, increasing, bounded self-dual operators which act additively on connected components. For that let us recall the definition and the essential properties of a saturation operator.

Proposition 3.38. *Let $T : \mathcal{B}(\overline{\Omega}) \to \mathcal{B}(\overline{\Omega})$ be a connected operator. Suppose that T is increasing, self-dual, bounded and acts additively on connected components. Let $Ker\,T := \{X \subseteq \overline{\Omega} : T(X) = \emptyset\}$. Let $X \subseteq \overline{\Omega}$ be an open or closed connected set. Then, if $\mathrm{Sat}(X) \notin Ker\,T$, we have*

$$T(X) = \mathrm{Sat}(X) \setminus \bigcup \{H : H \in \mathcal{H}(X) \text{ and } H \notin Ker\,T\}.$$

where $\mathcal{H}_c(X)$ denotes the family of internal holes of X.

Proof. Since T is self-dual, without loss of generality, we may assume that $p_\infty \notin X$. Since T is a connected operator, if Z is simply connected, then $T(Z)$ must be one of the sets $\{\emptyset, Z, \overline{\Omega} \setminus Z, \overline{\Omega}\}$. Since T is bounded we must have that either $T(Z) = \emptyset$ or $T(Z) = Z$. In particular, either $T(\mathrm{Sat}(X)) = \emptyset$ or $= \mathrm{Sat}(X)$. In the first case, we have that $\mathrm{Sat}(X) \in KerT$. In the second case, using the additivity of T on connected components and the observation at the beginning of the proof, we have

$$\begin{aligned} T(\overline{\Omega} \setminus X) &= T(\overline{\Omega} \setminus \mathrm{Sat}(X)) \bigcup \{T(H) : H \in \mathcal{H}_c(X)\} \\ &= (\overline{\Omega} \setminus \mathrm{Sat}(X)) \bigcup \{T(H) : H \in \mathcal{H}_c(X) \text{ and } H \notin Ker\,T\} \\ &= (\overline{\Omega} \setminus \mathrm{Sat}(X)) \bigcup \{H : H \in \mathcal{H}_c(X) \text{ and } H \notin Ker\,T\}. \end{aligned}$$

Now, using the self-duality of T we have

$$T(X) = \overline{\Omega} \setminus T(\overline{\Omega} \setminus X) = \mathrm{Sat}(X) \setminus \bigcup \{H : H \in \mathcal{H}_c(X) \text{ and } H \notin Ker\,T\}. \quad \square$$

Obviously, if T is increasing, then $Ker\,T$ is an ideal of sets, i.e., if $Y \subseteq X$ and $X \in Ker\,T$, then $Y \in Ker\,T$.

Theorem 3.39. *Let $T : \mathcal{B}(\overline{\Omega}) \to \mathcal{B}(\overline{\Omega})$ be an upper semicontinuous operator on closed sets which is increasing and self-dual. Then the associated operator $\tilde{T}$ is increasing and self-dual when defined on continuous functions.*

Proof. Since T is upper semicontinuous on closed sets we know that $\tilde{T}$ can be defined on upper semicontinuous functions, hence, on continuous functions. We are not saying that the image by $\tilde{T}$ of an upper semicontinuous function is upper semicontinuous. This would be true if the image by T of a closed set would be closed, and, then also the image by $\tilde{T}$ of a continuous function would be continuous.

Let $u \in C(\overline{\Omega})$. Since T is self-dual, we know that

$$T([u \geq \lambda]) = \overline{\Omega} \setminus T([u < \lambda]) \tag{3.19}$$

for all $\lambda \in \mathbb{R}$. These identities will imply that

$$\tilde{T}(-u) = -\tilde{T}(u). \tag{3.20}$$

Indeed, to prove (3.20) it will be sufficient that (3.19) holds for a dense subset of values of $\lambda \in \mathbb{R}$. Thus, let Λ be a dense subset of $\mathbb{R}$ such that (3.19) holds for any $\lambda \in \Lambda$. Let $x \in \overline{\Omega}$. We know that

$$\tilde{T}(u)(x) = \sup\{\lambda \in \Lambda : x \in T([u \geq \lambda])\}.$$

Using (3.19) we may write

$$\tilde{T}(u)(x) = \sup\{\lambda \in \Lambda : x \in \overline{\Omega} \setminus T([u < \lambda])\}.$$

Since $-\Lambda = \{-\lambda : \lambda \in \Lambda\}$ is also dense in $\mathbb{R}$, we have

$$\begin{aligned} -\tilde{T}(-u)(x) &= -\sup\{\mu' \in -\Lambda : x \in T([-u \geq \mu')\} \\ &= -\sup\{\mu' \in -\Lambda : x \in T([u \leq -\mu'])\} \\ &= -\sup\{\mu' \in -\Lambda : x \in T([u < -\mu'])\} \\ &= -\sup\{-\mu \in -\Lambda : x \in T([u < \mu])\} \\ &= \inf\{\mu \in \Lambda : x \in T([u < \mu])\}. \end{aligned}$$

The only identity of the above ones which needs a comment is the third. It is an immediate consequence of the increasing character of T and the density of Λ. Let us prove that

$$\inf\{\mu \in \Lambda : x \in T([u < \mu])\} = \sup\{\lambda \in \Lambda : x \in \overline{\Omega} \setminus T([u < \lambda])\}. \tag{3.21}$$

Let μ_i, λ_s be the left and right hand side of (3.21), respectively. Let $\lambda > \lambda_s$, $\lambda \in \Lambda$. Then $x \notin \overline{\Omega} \setminus T([u < \lambda])$, i.e., $x \in T([u < \lambda])$. Thus

$$(\lambda_s, \infty) \cap \Lambda \subseteq \{\mu \in \Lambda : x \in T([u < \mu])\}.$$

It follows that $\mu_i \leq \lambda_s$. Suppose that $\mu_i < \lambda_s$. Let $\mu_i < \mu < \rho < \lambda < \lambda_s$, $\mu, \rho, \lambda \in \Lambda$. Then $x \in T([u < \mu]) \subseteq T([u < \rho])$. On the other hand $x \in \overline{\Omega} \setminus T([u < \lambda])$. Since $\rho < \lambda$, also $x \in \overline{\Omega} \setminus T([u < \rho])$. This contradiction proves that $\mu_i = \lambda_s$. It follows that $\tilde{T}(-u)(x) = -\tilde{T}(u)(x)$. □

We have shown that G_ε is upper semicontinuous on closed sets and increasing (moreover, it acts additively on connected components and it is bounded). Thus, Theorem 3.39 would imply that $\tilde{G}_\varepsilon$ is a dual operator on $C(\overline{\Omega})$, if G_ε were self-dual. The fact is that G_ε is not self-dual. Indeed, let X be an open or closed ball in $\overline{\Omega}$ such that $|X| = \varepsilon$ and $p_\infty \notin X$. Then $G_\varepsilon X = X$ and $G_\varepsilon(\overline{\Omega} \setminus X) = \overline{\Omega}$, and we see that G_ε is not self-dual. The problem comes from connected sets which have saturations or holes of area ε. One could think that this could be avoided if we used the definition

$$G_\varepsilon X = \bigcup \{ \operatorname{Sat}(C) \setminus \cup_i C'_i : C \in \mathcal{CC}(X), |\operatorname{Sat}(C)| > \varepsilon, \\ C'_i \text{ internal hole of } C, |C'_i| > \varepsilon\}, \tag{3.22}$$

but in this case G_ε would not be upper semicontinuous. Indeed, let $X_n = \overline{B(p,r_n)} \subseteq \overline{\Omega}$ and $X = \overline{B(p,r)}$ with $r_n > r$, $r_n \downarrow r$ and $|X| = \varepsilon$. Then, using (3.22) we have $G_\varepsilon X_n = X_n$ while $G_\varepsilon X = \emptyset$. Thus, this definition is not satisfactory if we want to deduce easily that G_ε induces an operator acting on $\mathcal{USC}(\overline{\Omega})$. This is the reason why we choose our definition of G_ε given in (3.9). The inconvenience of G_ε not being self-dual is not a major one, since it turns out that it is generically self-dual. Indeed, the following proposition holds.

Proposition 3.40. *Assume that E is a closed set such that E and $\overline{\Omega} \setminus E$ have each a finite number of connected components. Assume that for any $C \in \mathcal{CC}(E)$, $\operatorname{Sat}(C)$ and the holes of C have measure $\neq \varepsilon$. Then $G_\varepsilon(E) = \overline{\Omega} \setminus G_\varepsilon(\overline{\Omega} \setminus E)$*

This is indeed sufficient to prove that $\tilde{G}_\varepsilon$ is self-dual on $C(\overline{\Omega})$. We shall not give the full details of the proof, having chosen to provide a different proof which we find more direct. However, let us give a sketch of this proof based on sets, assuming Proposition 3.40. Let $u \in C(\overline{\Omega})$ and let $v = M_\varepsilon^+ M_\varepsilon^- u$. Then one can prove that almost all upper level sets of v, $[v \geq \lambda]$, satisfy the assumptions of Proposition 3.40. We conclude that

$$G_\varepsilon[v \geq \lambda] = \overline{\Omega} \setminus G_\varepsilon[v < \lambda] \tag{3.23}$$

holds for almost all $\lambda \in \mathbb{R}$. Now, we recall that in Theorem 3.39 the identity

$$G_\varepsilon(-v) = -G_\varepsilon(v) \tag{3.24}$$

could be proved when (3.23) was true for a dense set of λ, and this is here the case. Thus (3.24) holds. Now, approximating u by $v = M_\varepsilon^+ M_\varepsilon^- u$, letting $\varepsilon \to 0+$ and using the continuity of $\tilde{G}_\varepsilon$ under uniform convergence we deduce that $\tilde{G}_\varepsilon(-u) = -\tilde{G}_\varepsilon(u)$. The full proof would require more space than the one we devoted here. Finally, let us note that the present proof is essentially based on Lemma 3.26, which also contributes to the proof of Theorem 3.27 and to the proof that $\tilde{G}_\varepsilon$ maps $C(\overline{\Omega})$ into $C(\overline{\Omega})$. This last result could be obtained differently, through approximations by $v = M_\varepsilon^+ M_\varepsilon^- u$, but the fact that $\tilde{G}_\varepsilon$ maps $\mathcal{USC}(\overline{\Omega})$ into $\mathcal{USC}(\overline{\Omega})$ requires the approach presented here.

Finally, let us notice that, when defined on $\mathbb{R}^N$, $\tilde{G}_\varepsilon$ is translation invariant, since the notions of connected set, holes, and measure are translation invariant.

3.5.3 *Interpretation*

Similar remarks to those for the extrema filters can be made concerning the shapes of $\tilde{G}_\varepsilon$, as defined in Chap. 2. In that chapter, the "shapes" of an image refer to the saturations of its connected components of level sets. The shapes of $\tilde{G}_\varepsilon u$ are the shapes of u of sufficient measure. $\tilde{G}_\varepsilon$ corresponds to a pruning of the tree of shapes of u.

3.6 Experiments

3.6.1 *Algorithm*

We propose to adapt the algorithm proposed by Vincent in [114, 115] to the self-dual grain filter. This algorithm is very close to the one proposed in [79, 80].

We scan the image pixel by pixel and modify the image, ensuring the following property: if $\mathcal{S}$ is the set of scanned pixels, the connected components of level sets without internal holes and meeting $\mathcal{S}$ have area (i.e., number of pixels) at least ε. Once all pixels are scanned, this ensures that the image is a fixed point of the grain filter.

At pixel P, we extract by a classical region growing algorithm the connected component of the isolevel set containing P, call it C and its level λ. There are different cases:

- Either C has neighbors with levels less and larger than λ, or C has an internal hole, or its area is at least ε. In this case, we do not modify the image and resume the scan.
- C has all neighbor pixels at level less (resp. larger) than λ, no internal hole, and its area is less than ε. Then C is a regional extremum. We put all pixels of C at the gray level of the neighbor pixel which is closest to λ and continue the region growing by extracting the connected component of the isolevel set of the modified image at this new level.

From these cases, it appears clearly that it is useful to initiate the region growing only at pixels which are local extrema of the gray level.

In order to memorize neighbors of the current region and their levels, it is convenient to use priority queues, as proposed in [114, 115], the priority of a pixel at level λ' being given by $-|\lambda - \lambda'|$. A convenient way to store this structure is to use a heap, as in a heap sort algorithm.

3.6.2 *Experimental Results*

Theoretically, the filters $M_\varepsilon^+ M_\varepsilon^-$, $M_\varepsilon^- M_\varepsilon^+$ and G_ε are different. This is illustrated in Fig. 3.2. The second row shows the filtered images $G_3 u$, $M_3^+ M_3^- u$

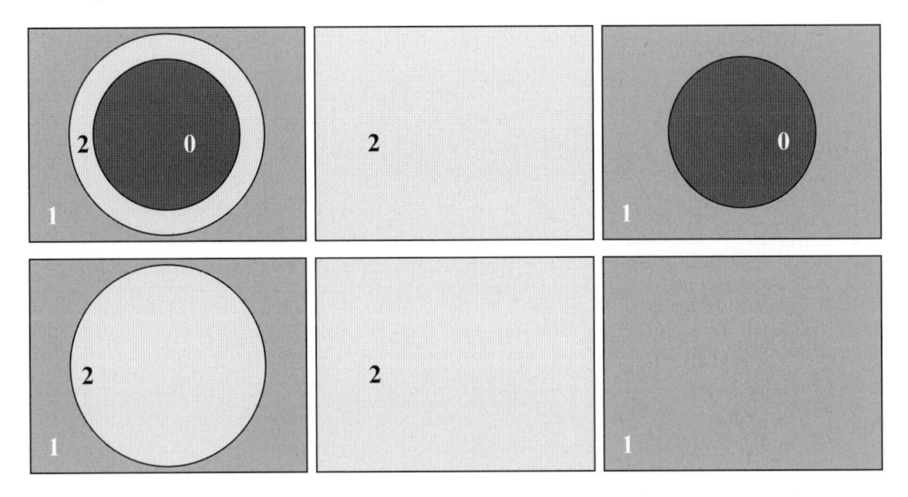

Fig. 3.2 Top-left: original image u. The three constant regions are supposed of area 2. Bottom left: G_3u. Middle column: M_3^-u and $M_3^+M_3^-u$. Right column: M_3^+u and $M_3^-M_3^+u$

Fig. 3.3 Texture image of a carpet, size 254 × 173

and $M_3^-M_3^+u$; they are all different, stressing that their respective notions of grains are different.

The difference appears in presence of holes. Concerning natural images, the difference would be most apparent on certain images of textures, for which the nestedness of shapes would be important. Figure 3.3 shows a complex texture, and Fig. 3.4 the image filtered according to the three filters of parameter 30

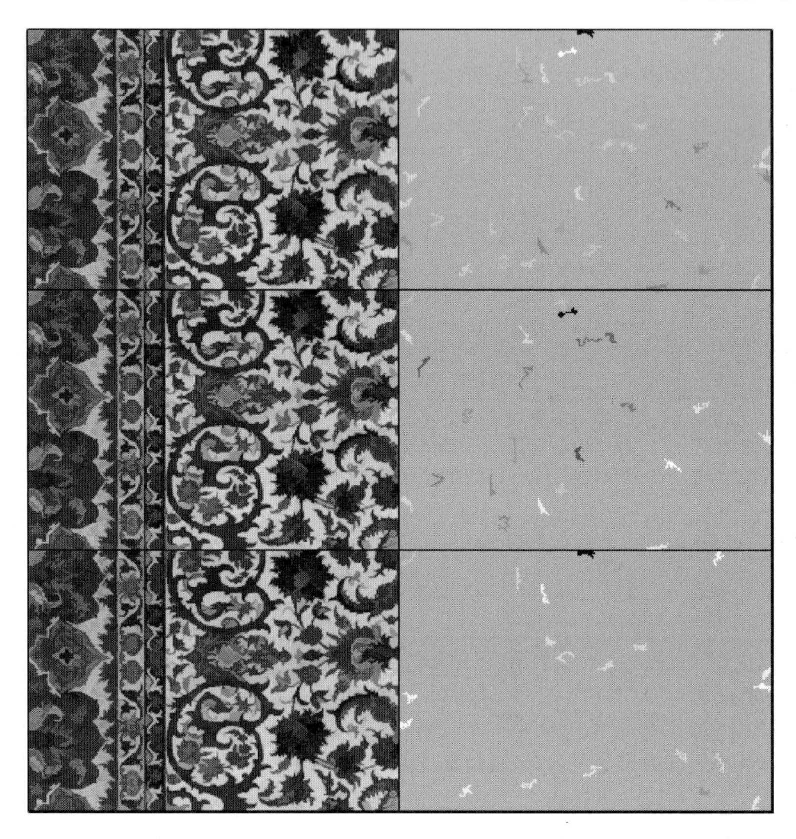

Fig. 3.4 Three grain filters applied to the image u of Fig. 3.3. Left column: $G_{30}u$, $M_{30}^{+}M_{30}^{-}u$ and $M_{30}^{-}M_{30}^{+}u$. Right column: difference images; $M_{30}^{+}M_{30}^{-}u - M_{30}^{-}M_{30}^{+}u$, $G_{30}u - M_{30}^{-}M_{30}^{+}u$ and $G_{30}u - M_{30}^{+}M_{30}^{-}u$

pixels. Whereas they are actually different, they are visually equivalent, and distinguishing them requires some effort. We can explain this by the fact that connected components of level sets having a hole of greater area than themselves are scarce. In other words, the situation illustrated by Fig. 3.2 is not frequent.

Finally, the effect of the grain filter with respect to impulse noise is illustrated in Fig. 3.5. The top-right image has got impulse noise of frequency 20%, meaning that around 1 pixel out of 5 is changed to an arbitrary value (independent of the original value). It is visible that after application of the grain filter G_{40}, the images are fairly close. The level lines become noisy, but for most of them, their presence or absence is not affected by noise.

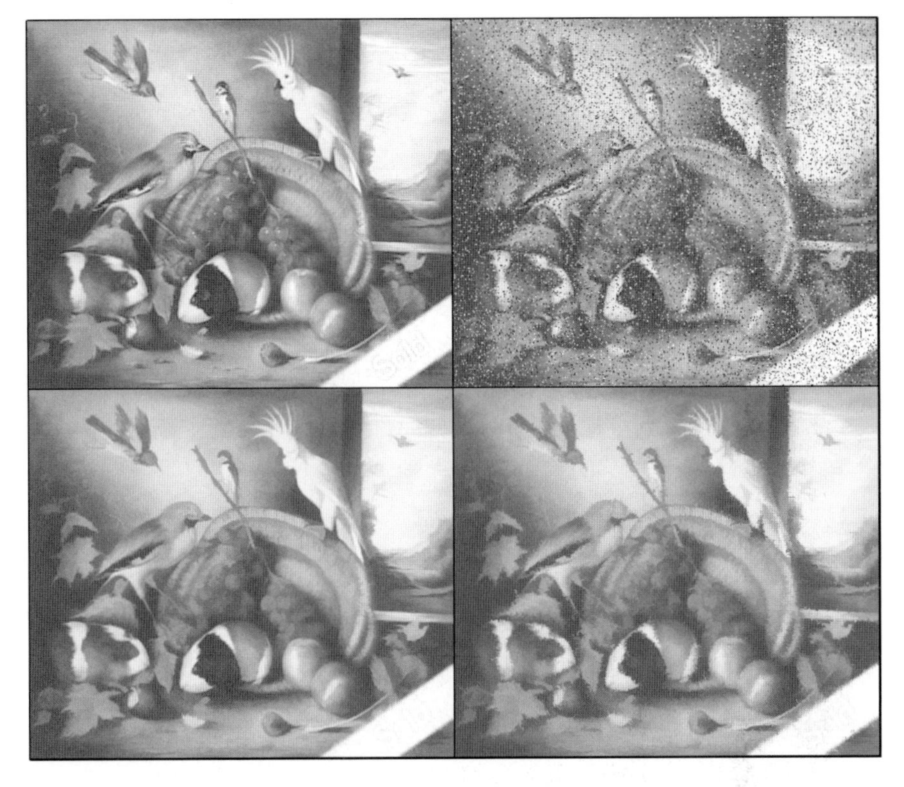

Fig. 3.5 Noise removal and topographic map preservation by the grain filter. Top-left: Original image (350 × 300). Top-right: Impulse noise of frequency 20%. Bottom row: grain filter G_{40} applied to these images

3.6.3 Complexity

The theoretical complexity of the algorithm is not easy to estimate, as it depends on image complexity. However, we can give experimental computation times. Measurements are done on a Pentium 2 processor at 300 MHz. For original image of Fig. 3.5, the computation time varies from 1.3 to 1.7 seconds, according to area parameter ε. For the same image with the noise of Fig. 3.5, whose topographic map is abnormally complex, these computations times vary from 1.7 to 2.2 seconds. This shows that reasonable times can be reached on rather weak processor speeds with respect to current standards.

Chapter 4
A Topological Description of the Topographic Map

The aim of Morse theory is to describe the topological changes of the (iso)level sets of a real valued function in terms of its critical points. Our purpose in this chapter is to describe several different notions of critical values (one of them will be called critical value, the other singular value) and prove that they are equivalent. One of those notions, the one of singular value, is intuitively related to the classical notion of critical value for a smooth function (points where the gradient vanishes), the other notion called critical value is based on a computational approach. When $N = 2$, this algorithm computes the Morse structure of the image from its upper and lower level sets. When $N \geq 3$, our notion of critical value is of topological nature and not equivalent to the notion of critical value used in classical Morse theory. Indeed, a function which embeds in its level sets the transformation of a sphere into a torus by pinching a hole into it may have no critical values (in our sense), besides a maximum. Due to that, we shall refer to it as a weak version of Morse theory of the topographic map. The results of this Chapter will be used in Chapter 5 in order to justify, in any dimension $N \geq 2$, the construction of the tree of shapes by merging of the trees of connected components of upper and lower level sets. Let us mention that when $N = 3$ the topological type of level surfaces could be later computed on the tree. Since this is not essential for our purposes, we shall not pursue it in the present text.

4.1 Monotone Sections and Singular Values

In this section we introduce the notions of monotone and maximal monotone sections of the topographic map which are the ones containing no topological changes of the topographic structure. As a consequence of the analysis of this chapter, the notions of monotone section of the tree and of the topographic map coincide. This gives a description of the topological structure of monotone sections of the tree of shapes.

As in previous chapters, let $\overline{\Omega}$ be a set homeomorphic to the closed unit ball of R^N ($N \geq 2$), $\{x \in R^N, \|x\| \leq 1\}$, and Ω be the interior of $\overline{\Omega}$. Note

V. Caselles and P. Monasse, *Geometric Description of Images as Topographic Maps*, Lecture Notes in Mathematics 1984,
DOI 10.1007/978-3-642-04611-7_4, © Springer-Verlag Berlin Heidelberg 2010

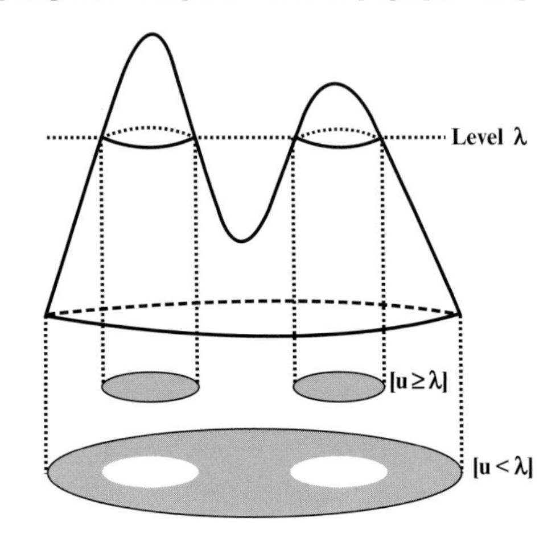

Fig. 4.1 A function u with upper and lower level sets.

that, in particular, $\overline{\Omega}$ is compact, connected and locally connected. Even though some of the results in this paper could be proved for more general sets, we shall assume that $\overline{\Omega}$ is of this form.

The (upper) topographic map of a function u is the family of the connected components of the level sets of u, $[u \geq \lambda]$, $\lambda \in \mathbb{R}$, the connected components being understood in the relative topology of $\overline{\Omega}$. It was proved in [18] that the topographic map is the structure of the image which is invariant under local contrast changes, a notion also defined in [18]. In [10, 11] the authors studied the topological weak Morse structure of the topographic map for continuous functions and upper semicontinuous functions. They defined the notion of monotone section as a region of the topographic map between two levels containing no topological change of the topographic structure, as it happens for smooth functions where singularities are understood in the usual way [75].

Let $u : \overline{\Omega} \to \mathbb{R}$ be a function. As usual, we shall denote $[u > \lambda] := \{x \in \overline{\Omega} : u(x) > \lambda\}$, $[u \leq \lambda] := \{x \in \overline{\Omega} : u(x) \leq \lambda\}$, $\lambda \in \mathbb{R}$, see Fig. 4.1. For each $\lambda, \mu \in \mathbb{R}$, $\lambda \leq \mu$ we define $[\lambda \leq u \leq \mu] = \{x \in \overline{\Omega} : \lambda \leq u(x) \leq \mu\}$, and we write $[u = \lambda] = [\lambda \leq u \leq \lambda]$. If I is an interval of $\mathbb{R}$ (be it open, closed, or half-open/half-closed), we denote $[u \in I] := \{x \in \overline{\Omega} : u(x) \in I\}$.

Definition 4.1. Let $u : \overline{\Omega} \to \mathbb{R}$ be a continuous function. A monotone section of the topographic map of u is a set of the form

$$X_I \in \mathcal{CC}([u \in I]), \tag{4.1}$$

where I is some interval of $\mathbb{R}$, such that for any $\alpha \in I$, the set $X_I \cap [u = \alpha]$ is a nonempty connected set.

Let us explain this definition. The notion of monotone section tries to capture a region of the topographic map of the image u where there is no change of topology, meaning no change in the number of connected components of its isolevel sets $[u = \alpha]$, $\alpha \in \mathbb{R}$. To get a better idea, let us assume that $u : \overline{\Omega} \to [a, b]$, $\overline{\Omega} \subseteq \mathbb{R}^2$, $a < b$, is smooth and each isolevel set $[u = \lambda]$ is a family of curves. The critical points of u are its maxima, minima and saddle points. If the function is smooth, we may identify these critical points using the usual rules of differential calculus. But there is also a topological description of them, which can be called its weak Morse description. We look at the isolevel sets $[u = \alpha]$ as α increases from a to b. Notice that if there is a minimum (resp. a maximum) at level α then a small curve appears (resp. disappears), and if there is a saddle point at level α there is some bifurcation in the curve, i.e., two curves merge, or a single curve splits into two (see Fig. 4.3.a). Thus, if we see the sets $[u = \alpha]$ as a family of moving curves, at the critical points, one of such curves appears, disappears, splits or merges. Then a connected components X of a set $[\lambda < u < \mu]$, $\lambda < \mu$, could be called monotone section if X contains no critical point of u (see Fig. 4.3.b). Notice that, if X is such a monotone section, then $X \cap [u = \alpha]$ is connected (a connected curve). Our definition of monotone section is inspired by this observation. Our purpose now is to find a partition of the image domain into its largest monotone sections, and the boundaries of these regions will be the curves corresponding to the critical values.

Proposition 4.2. *Let $u : \overline{\Omega} \to \mathbb{R}$ be a continuous function, and I be an interval of $\mathbb{R}$. If $X \in \mathcal{CC}([u \in I])$, then the following assertions are equivalent*

(i) X is a monotone section
(ii) for any subinterval $I' \subseteq I$, the set $X \cap [u \in I']$ is non-empty and connected
(iii) for any subinterval $I' \subseteq I$, the set $X \cap [u \in I']$ is non-empty and is connected component of $[u \in I']$.

Proof. Obviously, we have $(iii) \Leftrightarrow (ii) \Rightarrow (i)$. Let us prove that $(i) \Rightarrow (iii)$. Suppose that X is a monotone section but (iii) does not hold, that is, there is a subinterval $I' \subseteq I$ such that $X \cap [u \in I']$ is not connected. Notice that, since X is a monotone section, we have $u(X) = I$, and $u(X \cap [u \in I']) = I'$. Let A, B be two closed sets in $X \cap [u \in I']$ such that $A \cap B = \emptyset$ and $X \cap [u \in I'] = A \cup B$. Then

$$I' = u(A \cup B) = u(A) \cup u(B).$$

Observe that $u(A), u(B)$ are closed sets in I', and we deduce that $u(A) \cap u(B) \neq \emptyset$ because I' is connected. Let $\lambda \in u(A) \cap u(B)$. Then $A \cap [u = \lambda] \cap X \neq \emptyset$, $B \cap [u = \lambda] \cap X \neq \emptyset$, a contradiction, since $[u = \lambda] \cap X$ is connected. □

The following results permits us to define a monotone section which is maximal with respect to inclusion. Those sets are the non singular sets we mentioned above.

Lemma 4.3. *Let I_n be an increasing sequence of intervals and let $I = \cup_n I_n$. Let $X_n \in \mathcal{CC}([u \in I_n])$, $X_n \subseteq X_{n+1}$ for all n, and $X = \cup_n X_n$. Then $X \in \mathcal{CC}([u \in I])$. In particular, if X_n are monotone sections, so is X.*

Proof. Before going into the proof, let us observe that if K is a closed interval contained in I, then $K \subseteq I_n$ for n large enough. Now, since X_n is increasing, we may write $X_n = \mathrm{cc}([u \in I_n], p)$ for any $p \in X_1$. Obviously, $X = \cup_n X_n \subseteq \mathrm{cc}([u \in I], p)$. Suppose that $X \neq \mathrm{cc}([u \in I], p)$. Then there is $q \in \mathrm{cc}([u \in I], p) \setminus X$ and Γ a continuum joining p to q contained in $\mathrm{cc}([u \in I], p)$. Let $K = u(\Gamma) \subseteq I$. By our first observation, we know that $K \subseteq I_n$ for n large enough. This implies that $q \in X_n$ for n large enough, a contradiction. Hence $X = \mathrm{cc}([u \in I], p)$. It easily follows that X is a monotone section if X_n are. □

Proposition 4.4. *Assume that $u : \overline{\Omega} \to \mathbb{R}$ is a continuous function such that for each $\lambda, \mu \in \mathbb{R}$ with $\lambda \leq \mu$ the set $[\lambda \leq u \leq \mu]$ has a finite number of connected components. Let $I, (I_\alpha)_{\alpha \in A}$, be a family of intervals of $\mathbb{R}$. Assume that $X \in \mathcal{CC}([u \in I])$, $X_\alpha \in \mathcal{CC}([u \in I_\alpha])$ are monotone sections such that $X \cap X_\alpha \neq \emptyset$ for any $\alpha \in A$. Then $X \cup \bigcup_\alpha X_\alpha$ is a monotone section.*

Proof. Step 1. The case of two closed intervals. Let us consider first the case of two closed intervals I, J, and $X_I \in \mathcal{CC}([u \in I])$, $X_J \in \mathcal{CC}([u \in J])$ are monotone sections such that $X_I \cap X_J \neq \emptyset$ (hence $I \cap J \neq \emptyset$). First we will show that $X_I \cup X_J$ is a connected component of $[u \in I \cup J]$. Observe that $X_I \cup X_J$ is a connected set because is the union of intersecting connected sets, so $X_I \cup X_J \subseteq \mathcal{CC}([u \in I \cup J])$. To prove the other inclusion, we will proceed as follows. Let $V_{I \cup J}$ be the connected component of $[u \in I \cup J]$ containing $X_I \cup X_J$. Observe that X_I, X_J are closed sets in $\overline{\Omega}$. Let Z_I, Z_J be finite unions of connected components of $[u \in I]$, $[u \in J]$, respectively, such that $[u \in I] = X_I \cup Z_I$, $[u \in J] = X_J \cup Z_J$, $X_I \cap Z_I = \emptyset$, $X_J \cap Z_J = \emptyset$. Since Z_I is a closed set in $[u \in I]$ and this set is closed in $\overline{\Omega}$, Z_I is also a closed set in $\overline{\Omega}$. Similarly, Z_J is a closed set in $\overline{\Omega}$. Then the sets $Y_I = V_{I \cup J} \cap Z_I$ and $Y_J = V_{I \cup J} \cap Z_J$ are closed sets in $V_{I \cup J}$. Now, let us prove that $(X_I \cup X_J) \cap (Y_I \cup Y_J) = \emptyset$. Let us check that $Y_J \cap X_I = \emptyset$. Indeed $X_I \cap X_J \subseteq \{x \in X_I : u(x) \in J\} = \{x \in X_I : u(x) \in I \cap J\}$ the last set being a connected component Q of $[u \in I \cap J]$. In particular, we observe that $X_J \cap Q \neq \emptyset$ and we conclude that $Q \subseteq X_J$. Now, observe that $Y_J \cap X_I \subseteq Q$. Indeed, if $p \in Y_J \cap X_I$, then $p \in \{z \in X_I : u(x) \in J\} = Q$. Since $Q \subseteq X_J$, and $Y_J \subseteq Z_J$, this implies that $Y_J \cap X_I = \emptyset$. Similarly, $Y_I \cap X_J = \emptyset$. Since

$$V_{I \cup J} = (X_I \cup X_J) \cup (Y_I \cup Y_J)$$

and $V_{I \cup J}$ is connected then $Y_I = \emptyset$, $Y_J = \emptyset$. It follows that

$$V_{I \cup J} = X_I \cup X_J.$$

The proof that for any $\lambda \in I \cup J$, the set $V_{I\cup J} \cap [u = \lambda]$ is a connected component of $[u = \lambda]$ follows along the same lines of argument as the previous one.

Step 2. The case of any two intervals. Let I, J be any two intervals of $\mathbb{R}$ such that $X_I \in \mathcal{CC}([u \in I])$, $X_J \in \mathcal{CC}([u \in J])$ are monotone sections and $X_I \cap X_J \neq \emptyset$ (hence $I \cap J \neq \emptyset$). Then we may write $I = \cup_n I_n$, $J = \cup_n J_n$ where I_n, J_n are closed intervals of $\mathbb{R}$ and, by taking n large enough, if necessary, we may assume that $X_{I_n} \cap X_{J_n} \neq \emptyset$. By Step 1 we know that $X_{I_n} \cup X_{J_n}$ is a monotone section associated to the interval $I_n \cup J_n$, and, by Lemma 4.3, we deduce that $\cup_n (X_{I_n} \cup X_{J_n})$ is a monotone section associated to the interval $I \cup J$, and, indeed it coincides with $X_I \cup X_J$.

Step 3. The general case. Let X, X_α be as in the statement. By an iterated application of Step 2 we know that $X \bigcup \cup_{\alpha \in F} X_\alpha$ is a monotone section for any finite set $F \subseteq A$. Since, by Lindelöf's theorem, $X \bigcup \cup_{\alpha \in A} X_\alpha = X \bigcup \cup_{\alpha \in A_0} X_\alpha$ for a countable set $A_0 \subseteq A$, using Lemma 4.3 we conclude that $X \bigcup \cup_{\alpha \in A} X_\alpha$ is a monotone section. □

The previous Proposition is not true without the assumption
(**H**) *for any* $\lambda \leq \mu$ *the set* $[\lambda \leq u \leq \mu]$ *has a finite number of connected components.* Indeed, consider three circles A, B, C as in Fig. 4.2 where radius$(C) = 1$, and radius$(A) =$ radius$(B) + 1$. The function u takes the value 0 on $B \cup C$ and the value -1 on A. Let us take $u(x) = -d(x, B)$ in the region between the circles A and B and also outside A, $u(x) = d(x, B \cup C)$ when x is between the circles B and C, and $u(x) = d(x, C)\sin(1/d(x, C))$ if x is inside C. Observe that the region between A and B plus the circle C is a connected component of $[-1 \leq u \leq 0]$ which is a monotone section. On the other hand, the region between B and C is a connected component of $[0 \leq u \leq 1]$ which is also a monotone section. These two monotone sections intersect at $B \cup C$ and their union is not a monotone section. Indeed, it is only part of a connected component of $[-1 \leq u \leq 1]$, but not all, since the interior of C is missing.

As we shall prove in Proposition 4.23, any function with a finite number of regional extrema (see Definition 4.5) satisfies the assumption (**H**). Recall

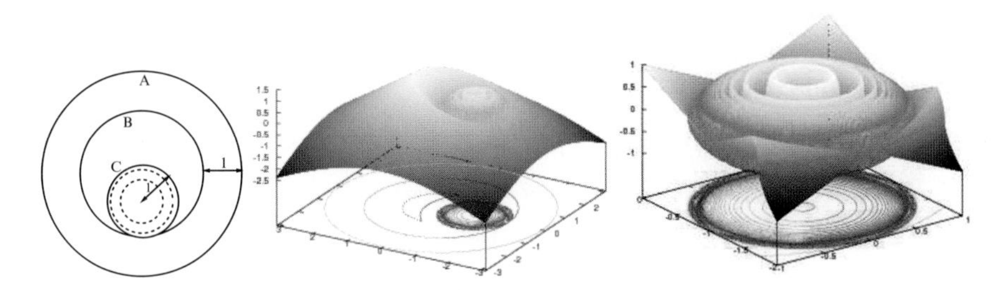

Fig. 4.2 Counterexample to Proposition 4.4 when hypothesis **H** is not verified. Left: the circles A, B and C (see text for details). Middle: some level sets of function u. Right: close-up around circle C.

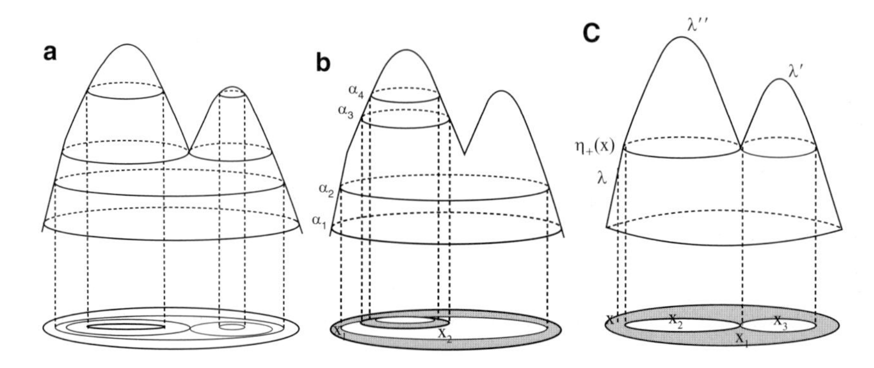

Fig. 4.3 **a**) As λ decreases and crosses a maximum a level curve appears. When crossing a saddle point, the isolevel curves $[u = \lambda]$ merge. **b**) The sets $[\alpha_1 \leq u \leq \alpha_2]$ and $[\alpha_3 \leq u \leq \alpha_4]$ are monotone sections. **c**) X_1,X_2,X_3 are three maximal monotone sections. X_1 is the maximal monotone section that contains the point x such that $u(x) = \lambda$, and it extends from $\eta_+(x)$ to $\eta_-(x)$ (we have not displayed this value). In this case, the isolevel set $[u = \eta_+(x)]$ is contained in X_1 but it is not contained in X_2 or X_3.

that, by Lemma 4.10 of Chap. 3, any function $u \in C(\overline{\Omega})$ such that $M_\delta^+ u = M_\delta^- u = u$ for some $\delta > 0$ has a finite number of regional extrema.

Assuming property (**H**), by Proposition 4.4, the union of monotone sections which intersect is a monotone section. This permits to define the notion of maximal monotone section containing a given point. Indeed, for any $x \in \overline{\Omega}$ the set $X_x = \mathcal{CC}([u = u(x)], x)$ (intuitively, the level curve through x) is a monotone section and, by Proposition 4.4, the set $\bigcup\{X : X$ is a monotone section containing $x\}$ is a *monotone section containing x which is maximal with respect to inclusion.* We call it the maximal monotone section containing x (see Fig. 4.3.c). It is associated to an interval $I(x)$ which may be open, closed, or half-open/half-closed. Obviously $u(x) \in I(x)$. Through the rest of this chapter we denote by $\eta_-(x), \eta_+(x)$ the end-points of the interval which may belong or not to $I(x)$.

Definition 4.5. Let $u \in C(\overline{\Omega})$ and $M \subseteq \overline{\Omega}$. We say that M is a regional maximum (resp., minimum) of u at height λ if M is a connected component of $[u = \lambda]$ and for all $\varepsilon > 0$ the set $[\lambda - \varepsilon < u \leq \lambda]$ (resp., $[\lambda \leq u < \lambda + \varepsilon]$) is a neighborhood of M.

Lemma 4.6. *Let $u \in \mathcal{C}(\bar{\Omega})$ be a function satisfying assumption* (**H**), *$\lambda \in \mathbb{R}$, $X \in \mathcal{CC}([u = \lambda])$ and $x \in X$. If X is a regional maximal of u, $\eta_+(x) = \lambda$. If X is a regional minimum of u, $\eta_-(x) = \lambda$.*

Proof. Let $\varepsilon > 0$ and suppose $Y = \mathrm{cc}([\lambda \leq u \leq \lambda + \varepsilon], X)$ is a monotone section of u. If X is a regional maximum, $[\lambda - \varepsilon < u \leq \lambda]$ is a neighborhood of X, so it contains an open set O including X. Then $O \cap Y \subseteq Y \cap [u = \lambda] = X$ since Y is a monotone section. On the other hand, $X \subseteq O \cap Y$, therefore

X is both open and closed in Y, showing that $Y = X$. This proves that $\eta_+(x) = \lambda$. A similar approach proves that $\eta_-(x) = \lambda$ when X is a regional minimum. □

Definition 4.7. Let $u \in C(\overline{\Omega})$ be a function satisfying assumption (**H**). We say that $\lambda \in \mathbb{R}$ is a singular value of u if it corresponds to a level where it begins or ends a maximal monotone section, i.e., there is a point $x \in \overline{\Omega}$ with $u(x) = \lambda$ such that $\eta_+(x) = \lambda$ or $\eta_-(x) = \lambda$.

By Lemma 4.6, regional maxima and minima of u are singular values of u.

Remark 4.8. Observe that the definition of singular value is self-dual, in the sense that, λ is a singular value of u if and only if $-\lambda$ is a singular value of $-u$.

4.2 Weakly Oscillating Functions, Their Intervals and the Structure of Their Topographic Map

Definition 4.9. We say that $u \in C(\overline{\Omega})$ is weakly oscillating if it has a finite number of regional extrema.

By Lemma 3.11 any function $u \in C(\overline{\Omega})$ such that $M_\delta^+ u = M_\delta^- u = u$ for some $\delta > 0$ is a weakly oscillating function.

In this section, unless explicitly stated, ∂ denotes the boundary operator in the relative topology of $\overline{\Omega}$.

Lemma 4.10. *Let $u \in C(\overline{\Omega})$ be a weakly oscillating function. Then for each $\lambda \in \mathbb{R}$, if $X \in \mathcal{CC}([u > \lambda])$ or $X \in \mathcal{CC}([u \geq \lambda])$ is nonempty, then X contains a regional maximum of u. A similar statement holds for lower level sets. Thus, for each $\lambda \in \mathbb{R}$, there is a finite number of connected components of $[u \geq \lambda]$ and each component has a finite number of holes.*

Proof. Let $X \in \mathcal{CC}([u > \lambda])$ be nonempty. Then $\mu := \max_{x \in X} u(x) > \lambda$ is attained at a point $p \in X$. Let $Y = \mathrm{cc}([u = \mu], p)$. Observe that $Y \subseteq X$. On the other hand, $u \leq \mu$ near Y, since otherwise we would find a point $q \in X$ with $u(q) > \mu$. We conclude that for all $\varepsilon > 0$ the set $[\mu - \varepsilon < u \leq \mu]$ is a neighborhood of Y, hence Y is a regional extremum of u. In particular, the number of connected components of $[u > \lambda]$ is finite.

Let us prove that each connected component of $[u \geq \lambda]$ contains a regional extremum of u. By the previous paragraph, we know that the connected components of $[u \geq \lambda]$ which intersect $[u > \lambda]$ contain a regional extremum and are finite in number. We denote them by $X_1, \ldots, X_k$. Let $X \in \mathcal{CC}([u \geq \lambda])$ be such that $X \subseteq [u = \lambda]$. Let us prove that X is a regional extremum of u. Obviously, X is a connected component of $[u = \lambda]$. Let $\eta > 0$ be such that $d(\cup_{i=1}^k X_i, X) \geq \eta$. We have that for all $\varepsilon > 0$ the set $[\lambda - \varepsilon < u \leq \lambda]$ is a neighborhood of X. Otherwise, there exists a sequence $p_n \to p \in X$ such that $u(p_n) > \lambda$. Then for each n, $p_n \in \cup_{i=1}^k X_i$ and, thus, $d(p_n, X) \geq \eta$, a contradiction since p_n converges to a point in X. We conclude that any connected component of $[u \geq \lambda]$ contains a regional maximum, and, thus, there must be a finite number of them.

Let $\lambda \in \mathbb{R}$. Let X be a connected component of $[u \geq \lambda]$ and let H be a hole of X. Observe that $\partial H \subseteq \partial X \subseteq X$. Since $X \cap \overline{H} \neq \emptyset$ and $X, \overline{H}$ are connected, then $X \cup H = X \cup \overline{H}$ is connected. If $H \subseteq [u \geq \lambda]$, then $H \subseteq X$, a contradiction. Hence $H \cap [u < \lambda] \neq \emptyset$. We conclude that each hole of X contains a component of $[u < \lambda]$. Hence there may be only a finite number of them. □

Lemma 4.11. *Let $u \in C(\overline{\Omega})$ be a weakly oscillating function. Let $X \in \mathcal{CC}([\lambda \leq u \leq \mu])$, $\lambda \leq \mu$, and let H be a hole of X. Then H is the saturation of a connected component either of $[u < \lambda]$ or of $[u > \mu]$.*

Proof. By Lemma 3.14 there exist a sequence of connected components $\{O_n\}_n$ of $[u < \lambda] \cup [u > \mu]$ such that $\mathrm{Sat}(O_n)$ are increasing and $H = \cup_n \mathrm{Sat}(O_n)$. Observe that O_n are two by two disjoint. Without loss of generality we may assume that O_n are all connected components of $[u < \lambda]$. Thus, by Lemma 4.10, there are only finitely many of them and there is a set $O \in \mathcal{CC}([u < \lambda])$ such that $H = \mathrm{Sat}(O)$. □

Lemma 4.12. *Let $u \in C(\overline{\Omega})$ be a weakly oscillating function. Let X be a connected component of $[\lambda \leq u \leq \mu]$, $\lambda \leq \mu$, and let L be a hole of X. Then there is some $\eta > 0$ such that either*

i) $\mathrm{Sat}(X, L) = \mathrm{Sat}(\mathrm{cc}([u \geq \lambda], X), L)$, *and $u < \lambda$ on $L_\eta := \{p \in L : d(p, X) < \eta\}$, or*
ii) $\mathrm{Sat}(X, L) = \mathrm{Sat}(\mathrm{cc}([u \leq \mu], X), L)$, *and $u > \mu$ on $L_\eta := \{p \in L : d(p, X) < \eta\}$.*

If the first case of the alternative holds, we say that L is a hole of negative type, in the second case we say that L is a hole of positive type.

Proof. We may assume that $L \neq \emptyset$, otherwise all saturations in the statement are equal to $\overline{\Omega}$ and the result is true. Assume that $\lambda < \mu$. By Lemma 4.11, we may write $L = \mathrm{Sat}(O)$ where either $O \in \mathcal{CC}([u < \lambda]$, or $O \in \mathcal{CC}([u > \mu])$. To fix ideas, assume that $O \in \mathcal{CC}([u < \lambda]$ (in particular, this implies that $[u < \lambda] \neq \emptyset$). Then $\partial L \subseteq \partial O \subseteq \partial[u < \lambda] \subseteq [u = \lambda]$. Let us prove that, for some $\eta > 0$, $u < \lambda$ on L_η.

Let us prove that the connected components of $[u \geq \lambda]$ are either disjoint to L, or contained in L. Let Y be a connected component of $[u \geq \lambda]$ intersecting L. Then $Y \subseteq L$. Otherwise, let $p \in Y \cap L$, $q \in Y \setminus L$, and let K be a continuum containing p and q and contained in Y. In this case, we have that $K \cap O \neq \emptyset$, a contradiction since $K \subseteq [u \geq \lambda]$. Indeed, if $K \cap O = \emptyset$, then K is contained in a hole of O. Since $\overline{\Omega} \setminus L$ is a hole of O containing $q \in K$, we have that $K \subseteq \overline{\Omega} \setminus L$, and this is a contradiction since $p \in K \cap L$.

Observe that if $Y \in \mathcal{CC}([u \geq \lambda])$, $Y \subseteq L$, then $\partial Y \cap \partial L = \emptyset$. Indeed, on one hand, we have $\text{dist}(Y, X) > 0$ (otherwise, if this distance is null, then $X \cap Y \neq \emptyset$, hence $X \subseteq Y \subseteq L$, a contradiction). This implies that $\partial Y \cap \partial X = \emptyset$. On the other hand, we have $\partial L \subseteq \partial X$. Hence, $\partial Y \cap \partial L = \emptyset$. Since, by Lemma 4.10, $[u \geq \lambda]$ has a finite number of connected components, we deduce that $\text{dist}([u \geq \lambda] \cap L, \partial L) > 0$, and, therefore, we have $L_\eta \subseteq [u < \lambda]$ for some $\eta > 0$. This implies that L is a hole of $\text{cc}([u \geq \lambda], X)$, and $\text{Sat}(X, L) = \text{Sat}(\text{cc}([u \geq \lambda], X), L)$.

Let us consider the case $\lambda = \mu$. By assumption X is a connected component of $[u = \lambda]$ and L is a hole of X. Let $y \in X$. Then $X = \cap_n X_n$ where $X_n = \text{cc}([\lambda \leq u \leq \lambda + \frac{1}{n}], y)$. Let $p \in L$. Then, by Lemma 2.13, we know that $\text{Sat}(X, p) = \cap_n \text{Sat}(X_n, p)$. Without loss of generality, we may assume that $p \notin X_n$ for all $n \geq 1$. But, according to the first part of the proof, we have that either $\text{Sat}(X_n, p) = \text{Sat}(\text{cc}([u \geq \lambda], y), p)$, or $\text{Sat}(X_n, p) = \text{Sat}(\text{cc}([u \leq \lambda + \frac{1}{n}], y), p)$. In the first case, we conclude that $\text{Sat}(X, p) = \text{Sat}(\text{cc}([u \geq \lambda], y), p)$. In the second case, using again Lemma 2.13, *ii*), we have that $\cap_n \text{Sat}(\text{cc}([u \leq \lambda + \frac{1}{n}], y), p) = \text{Sat}(\text{cc}([u \leq \lambda], y), p)$. Hence, $\text{Sat}(X, p) = \text{Sat}(\text{cc}([u \leq \lambda], y), p)$.

When $\text{Sat}(X, p) = \text{Sat}(\text{cc}([u \geq \lambda], y), p)$, L is a hole of $\text{cc}([u \geq \lambda], y)$. Hence $\partial L \subseteq \partial [u < \lambda]$ and the argument above proves that there is some $\eta > 0$ such that $u < \lambda$ on $L_\eta = \{p \in L : d(p, X) < \eta\}$. When $\text{Sat}(X, p) = \text{Sat}(\text{cc}([u \leq \lambda], y), p)$, L is a hole of $\text{cc}([u \leq \lambda], y)$. Then $\partial L \subseteq \partial [u > \lambda]$ and again the previous argument proves that there is some $\eta > 0$ such that $u > \lambda$ on $L_\eta = \{p \in L : d(p, X) < \eta\}$. □

A word of caution: when we say that X is a connected component of an upper or lower level set, we mean either that $X = \text{cc}([u \geq \lambda], p)$, or $X = \text{cc}([u > \lambda], p)$, or $X = \text{cc}([u \leq \lambda], p)$, or $X = \text{cc}([u < \lambda], p)$ for some $\lambda \in \mathbb{R}$, $p \in \overline{\Omega}$.

As a direct consequence of Lemma 4.12, we obtain:

Lemma 4.13. *Let $u \in C(\overline{\Omega})$ be a weakly oscillating function. Let $\lambda \leq \mu$. Let X be a connected component of $[\lambda \leq u \leq \mu]$, $\lambda \leq \mu$. Then* $\text{Sat}(X)$ *contains a connected component of an upper or lower level set of u.*

Definition 4.14. A sequence $A_1, \ldots, A_p$ of subsets of $\overline{\Omega}$ is called a chain if each A_i is contained in an internal hole of A_{i-1}, $i = 2, \ldots, p$.

Lemma 4.15. *Let $u \in C(\overline{\Omega})$ be a weakly oscillating function. Let X_n, $n \in \mathbb{N}$, be a connected component of $[\alpha_n \leq u \leq \beta_n]$, $\alpha_n \leq \beta_n$, such that $X_i \cap X_j = \emptyset$ for all $i \neq j$, $i, j \in \mathbb{N}$. Then (we may fix the point of infinity so that) there is an infinite chain formed by sets of the family X_n.*

Proof. First take an arbitrary point as point at infinity for saturation. According to Lemma 4.13, each $\text{Sat}\, X_n$ contains a regional extremum. As there are only a finite number of them, one regional extremum R is contained in infinitely many $\text{Sat}\, X_n$. After a possible extraction, we can suppose that

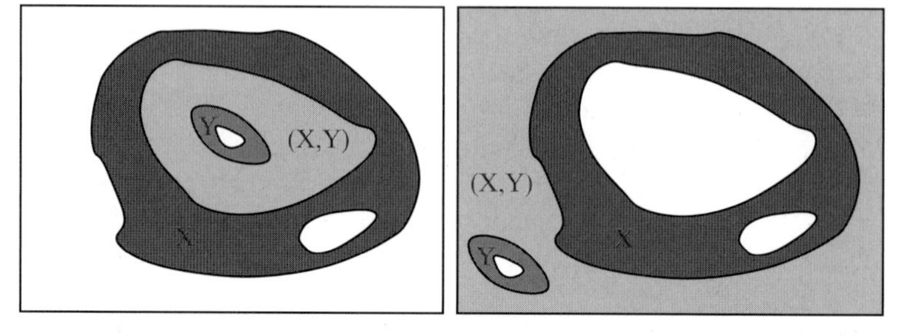

Fig. 4.4 Interval between two disjoint sets X and Y (dark gray). X has 3 holes and Y has 2. The interval (X,Y) is in light gray in both cases.

$R \subseteq \operatorname{Sat} X_n$ for all $n \in \mathbb{N}$. Then $\{\operatorname{Sat} X_n\}_n$ is totally ordered by inclusion, and, by extracting a subsequence, if necessary, we may assume either that $\operatorname{Sat} X_n$ is decreasing, or $\operatorname{Sat} X_n$ is increasing. In the first case, X_n is a chain. In the second case, we change our point of infinity to a point $p_\infty \in R$. With respect to this point, $\operatorname{Sat}(X_n, p_\infty)$ is decreasing and X_n forms a chain. □

Definition 4.16. Let X and Y be two disjoint subsets of $\overline{\Omega}$. We call intervals associated to X and Y the sets

$$(X,Y) = \overline{\Omega} \setminus (\operatorname{Sat}(X,Y) \cup \operatorname{Sat}(Y,X)),$$

$[X,Y) = (X,Y) \cup X$, $(X,Y] = (X,Y) \cup Y$ and $[X,Y] = (X,Y) \cup X \cup Y$.

To understand these definitions, we can choose p_∞ in X. If we note H_Y the hole of X including Y, we see that (X,Y) is $H_Y \setminus \operatorname{Sat} Y$. By symmetry, if we pick p_∞ in Y, then (X,Y) is $H_X \setminus \operatorname{Sat} X$, where H_X is the hole of Y containing X, see Fig. 4.4.

There is a close link between intervals and monotone sections. Essentially, if we discard the upper and lower levels, the picture that comes out is that monotone sections are like intervals containing no extrema. On the one hand, Proposition 4.20 gives sufficient conditions under which intervals associated to connected components of isolevels are monotone sections. On the other hand, as we shall prove in Proposition 4.21, a monotone section of a weakly oscillating function is an interval between two closed sets (not necessarily connected). Before stating and proving both Propositions, we prove a few simple lemmas that will be helpful.

Lemma 4.17. *Let X and Y be connected and closed disjoint subsets of $\bar{\Omega}$. Then*

(i) the intervals associated to X and Y are connected,
(ii) $\partial \operatorname{Sat}(X,Y), \partial \operatorname{Sat}(Y,X) \subseteq \partial(X,Y)$,
(iii) (X,Y) is open and $[X,Y]$ is closed.

Proof. (*i*) We can write $(X,Y) = A \cap B$ with $A = \bar{\Omega} \setminus \mathrm{Sat}(X,Y)$ and $B = \bar{\Omega} \setminus \mathrm{Sat}(Y,X)$. Then A and B are open and connected, and since $A \cup B = \bar{\Omega}$, which is unicoherent, (X,Y) is open and connected.
(*ii*) By symmetry, it suffices to prove one of the inclusions. Let $x \in \partial\,\mathrm{Sat}(X,Y)$. Since $(X,Y) \subseteq \bar{\Omega} \setminus \mathrm{Sat}(X,Y)$ and $B(x,r) \cap \mathrm{Sat}(X,Y) \neq \emptyset$ for any $r > 0$, we have $B(x,r) \cap (\bar{\Omega} \setminus (X,Y)) \neq \emptyset$ for any $r > 0$. On the other hand, since $\bar{\Omega} \setminus \mathrm{Sat}(Y,X)$ is a neighborhood of x, x is the limit point of a sequence $(x_n)_{n\in\mathbb{N}}$ in $(\bar{\Omega} \setminus \mathrm{Sat}(Y,X)) \setminus \mathrm{Sat}(X,Y) = (X,Y)$. We conclude that $\partial\,\mathrm{Sat}(X,Y) \subseteq \partial(X,Y)$.
(*iii*) Observe that, since $\partial\,\mathrm{Sat}(X,Y) \subseteq \partial X$, using (*ii*) we have that $\partial X \cap \partial(X,Y) \neq \emptyset$. Then, as X and (X,Y) are connected, we deduce that $[X,Y)$ is connected and for similar reasons $(X,Y]$ is also connected. If $(X,Y) \neq \emptyset$, $[X,Y) \cap (X,Y] = (X,Y) \neq \emptyset$, so their union $[X,Y]$ is also connected; if $(X,Y) = \emptyset$, then $X = Y = \emptyset$ and $[X,Y] = \emptyset$ is obviously connected. Finally, as $\partial(X,Y) \subseteq \partial\,\mathrm{Sat}(X,Y) \cup \partial\,\mathrm{Sat}(Y,X) \subseteq \partial X \cup \partial Y$, we have that

$$\partial[X,Y] \subseteq \partial(X,Y) \cup \partial X \cup \partial Y \subseteq \partial X \cup \partial Y \subseteq X \cup Y \subseteq [X,Y],$$

proving that $[X,Y]$ is closed. □

Lemma 4.18. *Let* $u \in C(\bar{\Omega})$.
(*i*) *Let* $X \subseteq \bar{\Omega}$. *If for some* $x \in X$, $u(x) \notin [\inf u(\partial X), \ \sup u(\partial X)]$, *then* X *contains a regional extremum. More precisely, if* $u(x) < \inf u(\partial X)$, *then* X *contains a regional minimum, and if* $u(x) > \sup u(\partial X)$, *then* X *contains a regional maximum.*
(*ii*) *Let* X *and* Y *be two different connected components of* $[u = \lambda]$. *Then there is a regional extremum in* (X,Y).

Proof. (*i*) Suppose $u(x) < \inf u(\partial X)$. $\bar{X}$ being a compact set, u reaches its minimum on $\bar{X}$ at some point y. Then $u(y) \leq u(x)$, thus $u(y) < \inf u(\partial X)$. Therefore $Y = \mathcal{CC}([u = u(y)], y)$ does not meet ∂X, and since Y is connected and meets X, Y is in the interior of X. Thus X is a neighborhood of Y and $u(y) = \min u(X)$, meaning that Y is a regional minimum.

The case $u(x) > \sup u(\partial X)$ can be dealt with in the same manner, by considering the maximum of u on $\bar{X}$.
(*ii*) The interval $[X,Y]$ being connected and containing X and Y, it is clear that $[X,Y] \not\subseteq [u = \lambda]$. There is thus some $x \in (X,Y)$ such that $u(x) \neq \lambda$. As $\partial(X,Y) \subseteq X \cup Y \subseteq [u = \lambda]$, using (*i*) we deduce that there is a regional extremum in (X,Y). □

Lemma 4.19. *Let* $u \in \mathcal{C}(\bar{\Omega})$ *be a weakly oscillating function,* $\lambda, \mu \in \mathbb{R}$, $\lambda < \mu$, $X \in \mathcal{CC}([u = \lambda])$, $Y \in \mathcal{CC}([u = \mu])$ *and assume that* (X,Y) *is a monotone section of* $[\lambda < u < \mu]$. *Then* (X,Y) *contains no regional extremum.*

Proof. Let $\alpha \in (\lambda, \mu)$ and $Z = (X,Y) \cap [u = \alpha]$, $z \in Z$. Since $z \in (X,Y)$, we have $\eta_-(z) \leq \lambda < \alpha$ and $\eta_+(z) \geq \mu > \alpha$. According to Lemma 4.6, we conclude that Z is not a regional extremum of u. □

Notice that this does *not* mean that any open monotone section contains no regional extremum. Indeed, we can have an open monotone section written $(X, Y]$ with Y a regional extremum.

Proposition 4.20. *Let $u \in C(\bar{\Omega})$ be a weakly oscillating function, $X \in \mathcal{CC}([u = \lambda])$ and $Y \in \mathcal{CC}([u = \mu])$ with $\lambda < \mu$. Suppose that (X, Y) contains no regional extremum. Then*

(i) *(X, Y) is a monotone section.*
(ii) *If all holes of X, except the one containing Y, are negative, $[X, Y)$ is a monotone section.*
(iii) *If all holes of Y, except the one containing X, are positive, $(X, Y]$ is a monotone section.*
(iv) *Under the assumptions of both (ii) and (iii), $[X, Y]$ is a monotone section.*

Proof. As *(iii)* can be deduced by applying *(ii)* to the function $-u$ instead of u, and as *(iv)* is a direct consequence of *(ii)* and *(iii)* by observing that $[X, Y] = [X, Y) \cup (X, Y]$ and $[X, Y) \cap (X, Y] = (X, Y) \neq \emptyset$, we shall prove only *(i)* and *(ii)*.

As (X, Y) is connected, $u\left((X, Y)\right)$ is an interval of $\mathbb{R}$. As $\partial(X, Y) \subseteq [u = \lambda] \cup [u = \mu]$, Lemma 4.18 proves that $u\left((X, Y)\right) \subseteq [\lambda, \mu]$. If $x \in [u = \lambda] \cap (X, Y)$, let $C = \mathcal{CC}([u = \lambda], x)$. Since $u(C) = \{\lambda\}$ we have $C \cap \partial(X, Y) = C \cap \partial X \subseteq C \cap X$, and since $X \neq C$ (otherwise we would have $X \cap (X, Y) \neq \emptyset$), we have $X \cap C = \emptyset$. We deduce that $C \cap \partial(X, Y) = \emptyset$, so that $C \subseteq (X, Y)$. Thus (X, Y) is a neighborhood of C and C is a regional minimum of u. This contradicts our hypotheses and proves that $\lambda \notin u\left((X, Y)\right)$. Likewise, the absence of a regional maximum in (X, Y) proves that $\mu \notin u\left((X, Y)\right)$, and therefore $(X, Y) \subseteq [\lambda < u < \mu]$.

Since $\partial(X, Y) \cap [\lambda < u < \mu] = \emptyset$, (X, Y) is open and closed in $[\lambda < u < \mu]$, and as it is connected, it is a connected component of $[\lambda < u < \mu]$.

We now prove that for any α such that $\lambda < \alpha < \mu$, $(X, Y) \cap [u = \alpha]$ is connected and not empty. Let $x \in \partial \operatorname{Sat}(X, Y)$. By Lemma 4.17, x is the limit point of a sequence $(x_n)_{n \in \mathbb{N}}$ in (X, Y). Since $u(x) = \lambda$, $\inf u(x_n) \leq \lambda$, so that $\inf u\left((X, Y)\right) \leq \lambda$. By a similar argument, $\sup u\left((X, Y)\right) \geq \mu$, and as $u\left((X, Y)\right)$ is a subinterval of $[\lambda < u < \mu]$, we get $u\left((X, Y)\right) = [\lambda < u < \mu]$ and therefore $(X, Y) \cap [u = \alpha] \neq \emptyset$ for $\lambda < \alpha < \mu$.

Suppose Z_1 and Z_2 are two distinct connected components of $(X, Y) \cap [u = \alpha]$. Since they do not intersect $[u = \alpha] \cap \partial(X, Y) = \emptyset$, it is clear that Z_1 and Z_2 are connected components of $[u = \alpha]$. Let us take saturations with respect to X. Then $\operatorname{Sat}(Z_1)$ and $\operatorname{Sat}(Z_2)$ contain a regional extremum. Since $\operatorname{Sat}(Z_i) \subseteq (X, Y) \cup \operatorname{Sat}(Y, X)$ and (X, Y) contain no regional extremum, we have $\operatorname{Sat}(Z_i) \cap \operatorname{Sat}(Y) \neq \emptyset$ implying $\operatorname{Sat}(Y) \subseteq \operatorname{Sat}(Z_i)$, for $i = 1, 2$. This implies that $\operatorname{Sat} Z_1$ and $\operatorname{Sat} Z_2$ are nested and it is easy to see that $(Z_1, Z_2) \subseteq (X, Y)$. But according to Lemma 4.18.*(ii)*, this implies the existence of a regional extremum included in (X, Y), which contradicts our assumptions. This ensures that $(X, Y) \cap [u = \alpha]$ has only one connected component, finishing the proof that (X, Y) is a monotone section.

Under the additional assumption of *(ii)*, we have $u([X, Y)) = u(X) \cup u((X, Y)) = [\lambda, \mu)$. Since $[X, Y) \cap [u = \lambda] = X$ is connected and for $\lambda < \alpha < \mu$, $[X, Y) \cap [u = \alpha] = (X, Y) \cap [u = \alpha]$ is connected and not empty, it remains only to show that $[X, Y)$ is a connected component of $[\lambda \leq u < \mu]$.

Let $H^1, \ldots, H^n$ the holes of X not containing Y, with $n \in \mathbb{N}$. For each $i \leq n$, there is some ε_i such that $H^i_{\varepsilon_i} \subseteq [u < \lambda]$. Let $\varepsilon = \min_i \varepsilon_i$. By taking the minimum with $d(X, Y)$, we can suppose that $\varepsilon \leq d(X, Y)$. The set $U = [X, Y) \bigcup_i H^i_\varepsilon$ is open since for $x \in X$, $B(x, \varepsilon)$ is in U. $U \cap [\lambda \leq u < \mu] = [X, Y)$, which shows that $[X, Y)$ is open in $[\lambda \leq u < \mu)$. Since its boundary in that set belongs to ∂X, which is in X, we see that $[X, Y)$ is also closed in $[\lambda \leq u < \mu]$, and since $[X, Y)$ is connected, it is a connected component of $[\lambda \leq u < \mu]$. □

Proposition 4.21. *Let $u \in C(\bar{\Omega})$ be a weakly oscillating function and C be a monotone section of u. Let us note $\lambda = \inf_C u$ and $\mu = \sup_C u$. Then $[u = \lambda] \cap \bar{C}$ and $[u = \mu] \cap \bar{C}$ are connected. Moreover, if we note $X = \mathcal{CC}([u = \lambda], [u = \lambda] \cap \bar{C})$ and $Y = \mathcal{CC}([u = \mu], [u = \mu] \cap \bar{C})$, then one of the four possibilities stands: $C = [X, Y]$, $C = [X, Y)$, $C = (X, Y]$, or $C = (X, Y)$.*

Proof. Since C is connected, $u(C)$ is an interval delimited by λ and μ. We will examine in turn the four possible intervals, each one will turn out to correspond to one choice in the alternative.

First, consider the case $u(C) = [\lambda, \mu]$. Then $C = \bar{C}$ and $C \cap [u = \lambda]$ and $C \cap [u = \mu]$ are connected as C is a monotone section. If $\lambda = \mu$, then $[X, Y] = [X, X] = X = C$. It remains to consider the case $\lambda < \mu$. Then $C \setminus X = C \cap [\lambda < u \leq \mu]$, which is connected and contains Y, so $C \setminus X \subseteq \bar{\Omega} \setminus \mathrm{Sat}(X, Y)$. In the same manner, $C \setminus Y \subseteq \bar{\Omega} \setminus \mathrm{Sat}(Y; X)$. By taking their intersection, $C \setminus (X \cup Y) \subseteq (X, Y)$. Since $\partial(C \setminus (X \cup Y)) \subseteq X \cup Y$, we deduce that $(X, Y) \cap \partial(C \setminus (X \cup Y)) = \emptyset$. Since (X, Y) is connected, it contains no nontrivial proper subset of empty boundary (for the topology of (X, Y)), thus $C \setminus (X \cup Y) = (X, Y)$, and we conclude that $C = [X, Y]$.

Next, consider the case $u(C) = [\lambda, \mu)$. Then $X = C \cap [u = \lambda]$ is connected. For $\alpha \in (\lambda, \mu)$, note $C_\alpha = C \cap [u = \alpha]$. Let $Z = \bigcap_{\alpha \in (\lambda, \mu)} \mathrm{Sat}(C_\alpha, X)$. Z is closed and connected, thanks to Zoretti's theorem ([51, §42,II,4]), and $\bar{\Omega} \setminus Z$ is also connected, so by unicoherency of $\bar{\Omega}$, ∂Z is connected. We now prove that $\partial Z = \bar{C} \cap [u = \mu]$.

Since $\bar{\Omega}$ is locally connected, we have

$$\partial Z \subseteq \overline{\bigcup_{\alpha \in (\lambda, \mu)} \partial(\mathrm{Sat}(C_\alpha, X))} \subseteq \overline{\bigcup_{\alpha \in (\lambda, \mu)} C_\alpha} \subseteq \bar{C}.$$

Let $\alpha \in (\lambda, \mu)$. Then $C \cap [\lambda \leq u \leq \alpha] = [X, C_\alpha]$ according to the first case, since it is a monotone section. So $[X, C_\alpha] \neq C$ and there is some $\beta \in (\alpha, \mu)$

such that $C_\beta \cap [X, C_\alpha] = \emptyset$. Thus $\mathrm{Sat}(C_\beta, X) \cap C_\alpha = \emptyset$. This yields $Z \cap C_\alpha = \emptyset$, and this being true for any $\alpha \in (\lambda, \mu)$, and recalling that $Z \cap C \cap [u = \lambda] = Z \cap X = \emptyset$, we have that $\partial Z \cap C \subseteq Z \cap C = \emptyset$, therefore $\partial Z \subseteq [u = \mu] \cap \partial C$. For the other inclusion, consider $x \in [u = \mu] \cap \partial C$ and $\alpha \in (\lambda, \mu)$. Since $x \notin [X, C_\alpha]$, we have $x \in \mathrm{Sat}(X, C_\alpha) \cup \mathrm{Sat}(C_\alpha, X)$. Suppose $x \in \mathrm{Sat}(X, C_\alpha)$, since $\partial\, \mathrm{Sat}(X, C_\alpha) \subseteq X$ and $x \notin X$, we can find a connected neighborhood U of x such that $U \subseteq \mathrm{Sat}(X, C_\alpha)$. Since $x \in \bar{C}$ and $x \notin X$, there is some $\beta \in (\lambda, \mu)$ such that $U \cap C_\beta \neq \emptyset$. Therefore $C_\beta \cap \mathrm{Sat}(X, C_\alpha) \neq \emptyset$. On the other hand $[C_\alpha, C_\beta]$ is connected and does not meet X, so $C_\beta \cap \mathrm{Sat}(X, C_\alpha) = \emptyset$. This contradiction proves that $x \in \mathrm{Sat}(C_\alpha, X)$. This being verified for any α, we get $x \in Z$. This shows that $[u = \mu] \cap \partial C \subseteq Z$ and clearly $\partial Z = [u = \mu] \cap \partial C$. We denote $Y = cc([u = \mu], \partial Z)$.

Since $C \cap Y = \emptyset$ and $X \subseteq C$, we have $C \subseteq \bar{\Omega} \setminus \mathrm{Sat}(Y, X)$. Furthermore, Y and $C \setminus X$ are connected and $\partial(C \setminus X) \cap Y \neq \emptyset$, so $Y \cup (C \setminus X)$ is connected and does not meet X, so $C \setminus X \subseteq \bar{\Omega} \setminus \mathrm{Sat}(X, Y)$. This shows that $C \setminus X \subseteq (X, Y)$. Since $\partial(C \setminus X) \subseteq X \cup Y$, $(X, Y) \cap \partial(C \setminus X) = \emptyset$ and as (X, Y) is connected, we have $C \setminus X = (X, Y)$ and $C = [X, Y)$.

Now, consider the case $u(C) = (\lambda, \mu]$. By applying the preceding case to $-u$ instead of u, we get easily $C = (X, Y]$.

Finally, consider the case $u(C) = (\lambda, \mu)$. Let $\alpha \in (\lambda, \mu)$. According to what precedes, $C \cap [\lambda < u \leq \alpha] = (X, C_\alpha]$ and $C \cap [\alpha \leq u < \mu] = [C_\alpha, Y)$. $[X, C_\alpha] \cup [C_\alpha, Y)$ being connected, we have

$$[X, C_\alpha] \cup [C_\alpha, Y) \subseteq \bar{\Omega} \setminus \mathrm{Sat}(Y, X)$$

and therefore $C \subseteq \bar{\Omega} \setminus \mathrm{Sat}(Y, X)$. In the same manner, $C \subseteq \bar{\Omega} \setminus \mathrm{Sat}(X, Y)$, so $C \subseteq (X, Y)$. As $\partial C \cap (X, Y) = \emptyset$ and (X, Y) is connected, we get $C = (X, Y)$. □

Proposition 4.22. *Let $u \in \mathcal{C}(\bar{\Omega})$ be a weakly oscillating function, $\lambda \in \mathbb{R}$, $X \in \mathcal{CC}([u = \lambda])$ and $x \in X$. Then*

(i) $\eta_+(x) > \lambda$ if and only if X has one, and only one, positive hole;
(ii) $\eta_-(x) < \lambda$ if and only if X has one, and only one, negative hole.

Proof. (ii) is a consequence of (i) applied to $-u$, so we prove only (i).

Suppose that $\eta_+(x) > \lambda$. According to Lemma 4.6, X is not a regional maximum, so X has at least one positive hole. There is some $\varepsilon > 0$ and $Y \in \mathcal{CC}([u = \lambda + \varepsilon])$ such that, according to Proposition 4.21, $[X, Y]$ is a monotone section. Obviously, Y is in a hole H of X and therefore $[X, Y] \subseteq X \cup H$. If we assume that X has at least two positive holes, there is a positive hole H' different from H. Then we can find some $\delta > 0$ such that H'_δ is in $[\lambda < u < \lambda + \varepsilon]$, and $H'_\delta \cup X$ is connected. Therefore $H'_\delta \cup X$ is in the connected component of $[\lambda \leq u \leq \lambda + \varepsilon]$ containing X, which is $[X, Y]$. This contradicts the fact that $H' \cap [X, Y] \subseteq H' \cap (X \cup H) = \emptyset$. We conclude that H is the unique positive hole of X.

Conversely, assume that X has a single positive hole H. Since there are a finite number of singular values, consider the smallest singular value μ larger than λ. We can find some $\delta > 0$ such that H_δ is in a connected component Z of $[\lambda < u < \mu]$. Since $X \cap \partial H_\delta \neq \emptyset$ and (λ, μ) has no singular value, Z is a monotone section, which, by Proposition 4.21, can be written as (X, Y) with $Y \in \mathcal{CC}([u = \mu])$ and $Y \subseteq H$. According to Lemma 4.19, (X, Y) has no extremum, so by application of the case (*ii*) of Proposition 4.20, $[X, Y)$ is a monotone section. Thus $\eta_+(x) \geq \mu > \lambda$. □

Proposition 4.23. *Let $u \in C(\overline{\Omega})$ be a weakly oscillating function. Then, for each $\lambda, \mu \in \mathbb{R}$ with $\lambda \leq \mu$, the set $[\lambda \leq u \leq \mu]$ has a finite number of connected components. In other words, u satisfies assumption* (**H**).

Proof. Suppose that $[\lambda \leq u \leq \mu]$ contains a countable family of connected components X_n. By Lemma 4.15, there is an infinite chain formed by sets of the family X_j. Now, using Lemma 4.18.(*ii*), we would obtain an infinite number of regional extrema in this chain. This contradiction proves that there is only a finite number of connected components of $[\lambda \leq u \leq \mu]$.

□

Remark 4.24. Observe that the assumption (**H**) on u does not imply that u is weakly oscillating. Indeed, let $x_n = \frac{1}{n}$, and $u : [0, 1] \to [0, 1]$ be such that $u(x_n) = x_n$, $u(0) = 0$, $u\left(\frac{x_n + x_{n+1}}{2}\right) = \frac{x_{n+1} + x_{n+2}}{2}$, u decreases linearly in $[x_{n+1}, \frac{x_n + x_{n+1}}{2}]$ and increases linearly in $[\frac{x_n + x_{n+1}}{2}, x_n]$. Then u satisfies assumption (**H**) but is not weakly oscillating.

Theorem 4.25. *Let $u \in C(\overline{\Omega})$ be a weakly oscillating function. Then there is a finite number of maximal monotone sections in the topographic map of u.*

Proof. Assume that there is a sequence $\{S_i\}_{i=1}^{\infty}$ of maximal monotone sections, each one associated with an interval (open, closed, halfopen/halfclosed) which will be denoted by $I_i = \{a_i, b_i\}$, $a_i \leq b_i$. Observe that, by Proposition 4.4, we have that $S_i \cap S_j = \emptyset$ for all $i \neq j$. Let $c_i \in I_i$, $i \geq 1$. Let $A_i = [u = c_i] \cap S_i$ which, by the definition of monotone section, is a connected component of $[u = c_i]$. Then, by Lemma 4.15, there is an infinite chain made of sets A_i. Thus we may assume that A_i form a chain with respect to a point p_∞ outside all of them. Observe that, since S_i are maximal monotone sections, we have that (A_{4i+1}, A_{4i+4}) is not a monotone section for any $i \geq 0$. Then, by Proposition 4.20, we know that each set (A_{4i+1}, A_{4i+4}), $i \geq 0$, contains a regional extremum of u, contradicting the fact that u is weakly oscillating. □

Finally, we characterize the limit shapes of a weakly oscillating function.

Proposition 4.26. *Let $u \in C(\overline{\Omega})$ be a weakly oscillating function. Then the limit shapes of u are sets of the form* $\mathrm{Sat}(C)$ *where either $C \in \mathcal{CC}([u \geq \lambda])$, or $C \in \mathcal{CC}([u > \lambda])$, or $C \in \mathcal{CC}([u \leq \lambda])$, or $C \in \mathcal{CC}([u < \lambda])$. Conversely, the sets of this form are limit shapes.*

Proof. Limit shapes are inf or sup operations on shapes. We may assume that these inf or sup operations are on a countable family of shapes. In the analysis of limit shapes of inf, resp. sup, type, we may assume that we are taking the intersection of a decreasing, resp. increasing, sequence of shapes S_n.

Let us analyze limit shapes of type $\cap_n S_n$. As we said above, we may assume that $S_n = \mathrm{Sat}(Q_n)$ is decreasing, and either all $Q_n \in \mathcal{CC}([u \geq \lambda_n])$, or all $Q_n \in \mathcal{CC}([u < \lambda_n])$. Assume first that $S_n = \mathrm{Sat}(Q_n)$, with $Q_n \in \mathcal{CC}([u \geq \lambda_n])$ for all n. Let us consider the following cases: *i)* If $Q_n \cap Q_{n+1} \neq \emptyset$ and $\lambda_n \leq \lambda_{n+1}$, then $Q_{n+1} \subseteq Q_n$. *ii)* If $Q_n \cap Q_{n+1} \neq \emptyset$ and $\lambda_n > \lambda_{n+1}$, then $Q_n \subseteq Q_{n+1}$. In this case $S_n \subseteq S_{n+1} \subseteq S_n$, i.e., $S_{n+1} = S_n$ and we may replace S_{n+1} by S_n. Thus, unless $S_m = S_n$ for all $m \geq n$, in which case $\cap_k S_k = S_n$, we may assume that we are always in case *i*). *iii)* If $Q_n \cap Q_{n+1} = \emptyset$ for a finite number of n, then we discard all of them and we start with an index n_0 such that *i*) or *ii*) holds for all $n \geq n_0$. Thus, we may assume that $Q_n \cap Q_{n+1} = \emptyset$ holds for countably many n. Then, by Lemma 4.10, u would have infinitely many regional maxima, a contradiction. This proves that we are always in case *i*) and we may assume that, if S_n are shapes of upper type, then (I) $S_n = \mathrm{Sat}(Q_n)$ with $Q_n \in \mathcal{CC}([u \geq \lambda_n])$ and λ_n increasing, so that Q_n is decreasing.

Assume now that $S_n = \mathrm{Sat}(Q_n)$ is decreasing, with $Q_n \in \mathcal{CC}([u < \lambda_n])$ for all n. As above, we consider the following cases: *i)* If $Q_n \cap Q_{n+1} \neq \emptyset$ and $\lambda_n \geq \lambda_{n+1}$, then $Q_{n+1} \subseteq Q_n$. *ii)* If $Q_n \cap Q_{n+1} \neq \emptyset$ and $\lambda_n < \lambda_{n+1}$, then $Q_n \subseteq Q_{n+1}$. Hence $S_n \subseteq S_{n+1} \subseteq S_n$, i.e., $S_{n+1} = S_n$ and we may replace S_{n+1} by S_n. Thus, unless $S_m = S_n$ for all $m \geq n$, in which case $\cap_k S_k = S_n$, we may assume that we are always in case *i*). *iii)* If $Q_n \cap Q_{n+1} = \emptyset$, for a finite number of n, then we discard all of them and we start with an index n_0 such that *i*) or *ii*) holds for all $n \geq n_0$. Thus, we may assume that $Q_n \cap Q_{n+1} = \emptyset$ holds for countably many n. Thus we are in the same situation as in case *iii*) of the above paragraph and, by Lemma 4.10, we would get infinitely many regional minima. This proves that we are always in case *i*) and we may assume that, if S_n are lower shapes, then (II) $S_n = \mathrm{Sat}(Q_n)$ with $Q_n \in \mathcal{CC}([u < \lambda_n])$ and λ_n decreasing, so that Q_n is decreasing.

With a similar argument we prove that to analyze the limit shape $\cup_n S_n$ we may assume that $S_n = \mathrm{Sat}(Q_n)$ are increasing and either (III) $Q_n \in \mathcal{CC}([u < \lambda_n])$ with λ_n increasing, so that Q_n is increasing, or (IV) $Q_n \in \mathcal{CC}([u \geq \lambda_n])$ with λ_n decreasing, so that Q_n is increasing.

Case (I): $S_n = \mathrm{Sat}(Q_n)$, $Q_n \in \mathcal{CC}([u \geq \lambda_n])$ with λ_n increasing, so that Q_n is decreasing. Since $\cap_n Q_n = Q$ where $Q \in \mathcal{CC}([u \geq \lambda])$, by Lemma 2.13, we have that

$$\cap_n S_n = \cap_n \mathrm{Sat}(Q_n) = \mathrm{Sat}(Q).$$

Case (II): $S_n = \mathrm{Sat}(Q_n)$, $Q_n \in \mathcal{CC}([u < \lambda_n])$ with λ_n decreasing, so that Q_n is decreasing. Observe that, without loss of generality, we may assume that λ_n is strictly decreasing. In this case, there is $Q'_n \in \mathcal{CC}([u \leq \lambda_n])$ such that $Q_n \subseteq Q'_n \subseteq Q'_{n-1}$, and we have

$$\cap_n Q_n = \cap_n Q'_n = Q,$$

for some $Q \in \mathcal{CC}([u \leq \lambda])$. Again, using Lemma 2.13 we get that

$$\cap_n S_n = \cap_n \mathrm{Sat}(Q'_n) = \mathrm{Sat}(Q).$$

Case (III): $S_n = \mathrm{Sat}(Q_n)$ where $Q_n \in \mathcal{CC}([u < \lambda_n])$ with λ_n increasing, so that Q_n is increasing. Let $p \in Q_1$. Then $Q_n = \mathrm{cc}([u < \lambda_n], p)$. Let $\lambda = \sup_n \lambda_n$. By Lemma 4.3 we have that $\cup_n Q_n = \mathrm{cc}([u < \lambda], p)$ and Lemma 2.13.(ii) implies that

$$\cup_n \mathrm{Sat}(\mathrm{cc}([u < \lambda_n], p)) = \mathrm{Sat}(\mathrm{cc}([u < \lambda], p). \tag{4.2}$$

Case (IV): $S_n = \mathrm{Sat}(Q_n)$ where $Q_n \in \mathcal{CC}([u \geq \lambda_n])$ with λ_n decreasing, so that Q_n is increasing. Without loss of generality, we may assume that λ_n are strictly decreasing. Let $\lambda = \inf_n \lambda_n$. Let $p \in Q_1$. Then $Q_n = \mathrm{cc}([u \geq \lambda_n], p)$. By Lemma 4.3 we have that $\cup_n Q_n = \cup_n \mathrm{cc}([u > \lambda_n], p) = \mathrm{cc}([u > \lambda], p)$, and, by Lemma 2.13.$(ii)$ we have

$$\cup_n \mathrm{Sat}(Q_n) = \mathrm{Sat}(\mathrm{cc}([u > \lambda], p)).$$

The last assertion is easily proved with the same techniques. □

4.3 Signature and Critical Values

This section is devoted to the definition of critical levels (or critical values). Before going into the formal definition, let us explain the idea behind it. Intuitively, critical levels are levels where a connected component of an isolevel set $[u = \alpha]$ appears, disappears, splits, or merges. Those critical levels can be identified by looking at $[u = \alpha]$ as the common boundary of two sets $[u \geq \alpha]$ and $[u < \alpha]$, and describing the topology of $[u = \alpha]$ in terms of the connected components of the sets $[u \geq \alpha]$ and $[u < \alpha]$. Indeed, if we increase α and we cross a level with a minimum (resp. maximum), then a connected component of $[u < \alpha]$ appears (resp. a connected component of $[u \geq \alpha]$ disappears). If we cross a saddle point then either two connected components of $[u < \alpha]$, or of $[u \geq \alpha]$, merge. This will give us a simple way to compute the critical levels. Then we shall prove the result that both notions of critical and singular values are equivalent, and this will prove that we are indeed computing the maximal monotone sections of the image.

Throughout the rest of this chapter we shall assume that $u \in C(\overline{\Omega})$ is weakly oscillating. We recall that, by Lemma 4.10, each level set has a finite number of connected components and a finite number of holes. Let $\mathcal{E}$ denote the set of regional extrema of u.

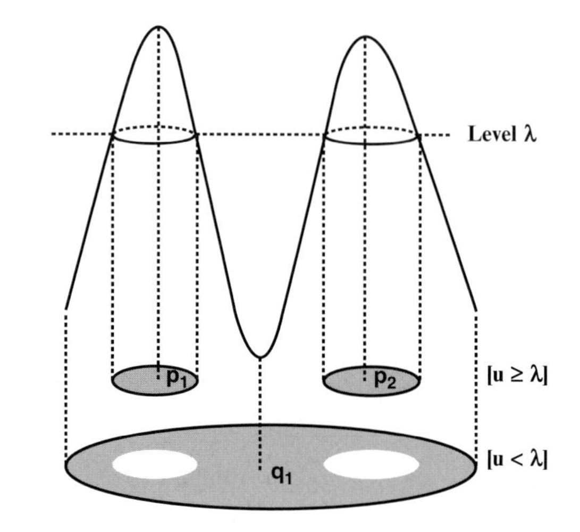

Fig. 4.5 A function u and its upper and lower level sets at level λ with its assigned signature. The set $[u \geq \lambda]$ has two connected components depicted in gray and its signature consists of two points $\{p_1, p_2\}$. The set $[u < \lambda]$ has only one connected component with two holes, and is depicted as the circular region below with the two white holes. Its signature consists of the point q_1.

Let us introduce the notion of critical level of u. For that we first need the following definition.

Definition 4.27. For $X \in \bar{\Omega}$, we note $\mathcal{E}(X)$ the set $\{E \in \mathcal{E} | E \subseteq X\}$. We define the signature of X as $sig(X) = \{\mathcal{E}(C) | C \in \mathcal{CC}(X)\}$. For $\lambda \in \mathbb{R}$, we define the signature of u at level λ the set $sig(\lambda) = sig([u \geq \lambda]) \cup sig([u < \lambda])$.

Notice that $sig(X)$ and $sig(\lambda)$ are in $\mathcal{P}(\mathcal{P}(\mathcal{E}))$. We remark that for any $\lambda \in (\min u, \max u]$, $sig(\lambda)$ is a partition of $\mathcal{E}$. That all elements of $sig(\lambda)$ are nonempty is a consequence of Lemma 4.10. Moreover, since $\mathcal{CC}([u \geq \lambda]) \cup \mathcal{CC}([u < \lambda])$ is a partition of $\bar{\Omega}$ and each connected component of an isolevel is contained in one element of this partition, if $E \in \mathcal{E}$, E is contained in one unique element C of this partition, therefore $E \in sig(C)$, and E belongs to no other element of $sig(\lambda)$.

The definition of signature is illustrated in Fig. 4.5. Note that in the case presented in this figure, $sig([u \geq \lambda]) = \{\{p_1\}, \{p_2\}\}$, $sig([u < \lambda]) = \{\{q_1\}\}$ and $sig(\lambda) = \{\{p_1\}, \{p_2\}, \{q_1\}\}$.

Our next lemma proves that for weakly oscillating functions the signature may only change from above. This justifies the definition of critical value that follows it.

Lemma 4.28. *Let $u \in C(\overline{\Omega})$ be a weakly oscillating function. Let $\lambda \in \mathbb{R}$. There is $\varepsilon > 0$ such that $sig(\mu)$ is constant for all $\mu \in (\lambda - \varepsilon, \lambda]$.*

Proof. Let $X^{\lambda,i}$, $X_{\lambda,j}$, $i = 1, \ldots, r$, $j = 1, \ldots, s$, be the family of connected components of $[u \geq \lambda]$, resp. $[u < \lambda]$. Let $i \in \{1, \ldots, r\}$. For each $\mu < \lambda$, let $X^{\mu,i}$ be the connected component of $[u \geq \mu]$ containing $X^{\lambda,i}$. Then, obviously, we have

$$X^{\lambda,i} \subseteq \cap_{\mu<\lambda} X^{\mu,i}.$$

Now, since $X^{\mu,i}$ is a decreasing sequence of continua their intersection is also a continuum [51]. Moreover, it is contained in $[u \geq \lambda]$. Therefore,

$$\cap_{\mu<\lambda} X^{\mu,i} \subseteq \mathrm{cc}([u \geq \lambda], p_i) = X^{\lambda,i},$$

and we have the equality of both sets. As a consequence, there is an $\varepsilon > 0$ such that for each $\mu \in (\lambda - \varepsilon, \lambda]$, the sets $X^{\lambda,i}$, $i = 1, \ldots, r$, are contained in different connected components of $[u \geq \mu]$. Moreover, since the number of connected components of each $[u \geq \mu]$ is finite, we may choose $\varepsilon > 0$ such that for each $\mu \in (\lambda - \varepsilon, \lambda]$ the set $[u \geq \mu]$ consists of r connected components, each one of them containing a different component of $[u \geq \lambda]$. Since u is weakly oscillating, for $\epsilon > 0$ small enough, the regional extrema of u in each $X^{\mu,i}$, $i = 1, \ldots, r$, is constant for $\mu \in (\lambda - \epsilon, \lambda]$.

Let $\mu_n \uparrow \lambda$. Again, using that $\cup_n [u < \mu_n] = [u < \lambda]$, for n large enough, we have that $[u < \mu_n] \cap X_{\lambda,j}$, $j = 1, \ldots, s$, are the connected components of $[u < \mu_n]$. As above, we know that the regional extrema of u in each $[u < \mu_n] \cap X_{\lambda,j}$ coincide with the regional extrema in $X_{\lambda,j}$, $j = 1, \ldots, s$, for n large enough. We conclude that there is an $\varepsilon > 0$ such that $sig(\mu)$ is constant for each $\mu \in (\lambda - \varepsilon, \lambda]$. □

Definition 4.29. Let $u \in C(\overline{\Omega})$ be a weakly oscillating function. We say that $\lambda \in \mathbb{R}$ is a critical value for u if there is a sequence $\mu_n \downarrow \lambda$ such that $sig(\mu_n) \neq sig(\lambda)$ for each $n = 1, 2, \ldots$

Let us explain the phenomena reflected by a change of signature. For simplicity, let us explain them at the discrete level. For that, let us consider two consecutive levels λ and $\lambda + 1$. Let $\cup sig([u \geq \mu]) = \cup_{C \in CC([u \geq \mu])} sig(C)$, $\mu \in \mathbb{R}$. Notice that $\cup sig([u \geq \lambda + 1]) \subseteq \cup sig([u \geq \lambda])$. If $sig(\lambda) \neq sig(\lambda + 1)$ several things may happen: (a) $sig([u \geq \lambda]) \neq sig([u \geq \lambda + 1])$ while $\cup sig([u \geq \lambda]) = \cup sig([u \geq \lambda + 1])$, (b) $sig([u < \lambda]) \neq sig([u < \lambda + 1])$ while $\cup sig([u < \lambda]) = \cup sig([u < \lambda + 1])$, (c) $\cup sig([u \geq \lambda]) \neq \cup sig([u \geq \lambda + 1])$ (hence $sig([u \geq \lambda]) \neq sig([u \geq \lambda + 1])$). In this last case we also have that $\cup sig([u < \lambda]) \neq \cup sig([u < \lambda + 1])$. If we are in case (a) there must be two connected components of $[u \geq \lambda + 1]$ that have merged at level λ and the corresponding signatures fused. If we are in case (b) there must be two connected components of $[u < \lambda + 1]$ that have split at level λ and the corresponding signatures split. If we are in case (c) then a regional extremum has been transferred from $\cup sig([u < \lambda + 1])$ to $\cup sig([u \geq \lambda])$ be either a regional maximum because a connected component of $[u \geq \lambda]$ appeared which was not present at level $\lambda + 1$, or a regional minimum because a connected component of $[u < \lambda + 1]$ disappeared at level λ. These two last cases could happen combined with merging or splitting of connected components (see Fig. 4.6).

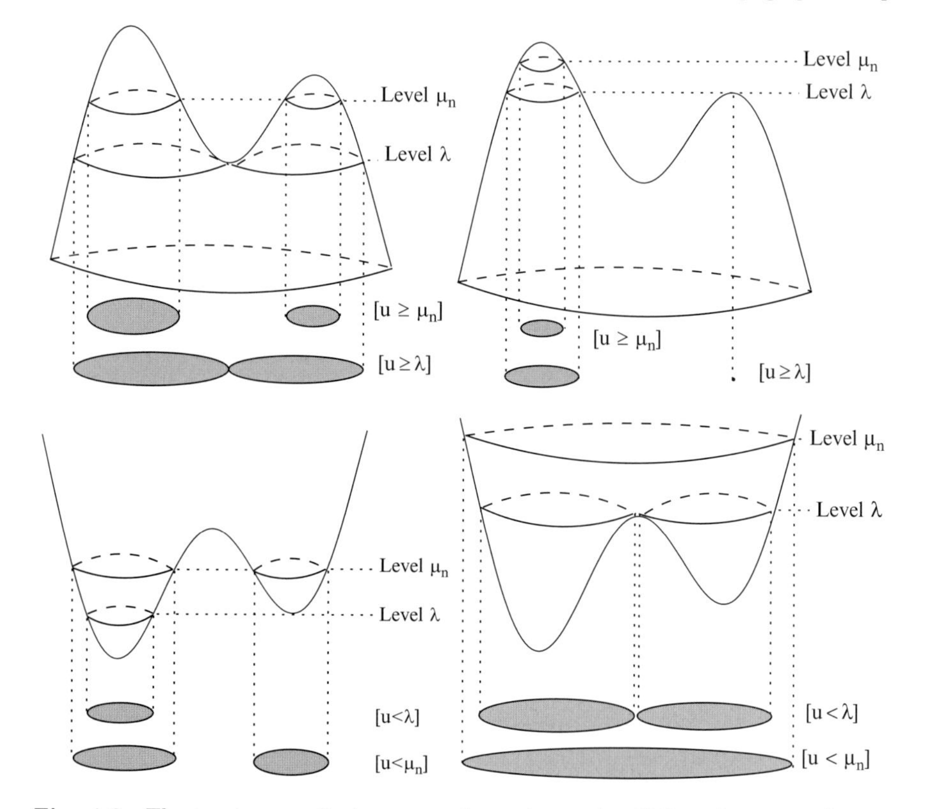

Fig. 4.6 The top images display cases of merging and splitting of connected components as λ decreases. The bottom figures display the birth of an upper connected component or the death of a lower one. Those are the changes of signature as λ varies.

Finally, observe that since the signature of any connected component of $[u \geq \lambda]$ increases (respectively the signature of any connected component of $[u < \lambda]$ decreases) as λ decreases, then there are only finitely many possible changes in $sig(\lambda)$. Thus, if $u \in C(\overline{\Omega})$ is weakly oscillating, then the number of critical values of u is finite. In particular, the signature $sig(\mu)$ is locally constant at each side of a critical value, i.e., if λ is a critical value, then there is $\varepsilon > 0$ such that

$$sig(\mu) = sig(\lambda) \neq sig(\mu') \quad \text{and } sig(\mu') \text{ is constant}$$

for each $\mu < \lambda < \mu'$, $\mu \in (\lambda - \varepsilon, \lambda)$, $\mu' \in (\lambda, \lambda + \varepsilon)$. This implies that the previous description of the changes of the topology of level sets for discrete images also holds in the continuous case.

4.4 Critical Versus Singular Values

Theorem 4.30. *Let $u \in C(\overline{\Omega})$ be a weakly oscillating function. Let $\lambda \in \mathbb{R}$. Then λ is a critical level of u if and only if λ is a singular level of the topographic map of u.*

Proposition 4.31. *Let $u \in C(\overline{\Omega})$ be a weakly oscillating function. If $\lambda \in \mathbb{R}$ is a critical value, then λ is also a singular value.*

Proof. Assume that λ is not a singular value of u. Then there is $\epsilon > 0$ such that there are no singular values of u in $(\lambda-\varepsilon, \lambda+\varepsilon)$, thus, in particular, the connected components of the set $[\lambda - \varepsilon < u < \lambda + \varepsilon]$ are monotone sections. We denote them by $M_1, \ldots, M_r$, and we observe that $u(M_i) = (\lambda - \varepsilon, \lambda + \varepsilon)$ for each $i = 1, \ldots, r$. Let $\lambda < \mu < \lambda + \varepsilon$. We observe that the connected components of $[u = \lambda]$, resp. $[u = \mu]$, are $M_i \cap [u = \lambda]$, resp. $M_i \cap [u = \mu]$, $i = 1, \ldots, r$. Then each connected component of $[u \geq \lambda]$ (there are r of them) may be written as the union of a connected component of $[u \geq \mu]$ and one of the sets $M_i \cap [\lambda \leq u \leq \mu]$ (for some $i \in \{1, \ldots, r\}$).

In a similar way, we have that any connected component of $[u < \mu]$ (there are r of them) may be written as the union of a connected component of $[u < \lambda]$ and one of the sets $M_i \cap [\lambda < u \leq \mu]$ (for some $i \in \{1, \ldots, r\}$). We observe that there are no regional extrema in the sets $M_i \cap [\lambda \leq u \leq \mu]$, $M_i \cap [\lambda < u \leq \mu]$, $i \in \{1, \ldots, r\}$.

We conclude that $sig(\lambda) = sig(\mu)$ for any $\mu \in (\lambda, \lambda + \varepsilon)$, hence λ is not a critical value of u. □

Proposition 4.32. *Let u be a weakly oscillating function and $\lambda, \mu \in \mathbb{R}$ such that $\min_{\overline{\Omega}} u \leq \lambda < \mu \leq \max_{\overline{\Omega}} u$. We assume that $sig(\lambda) = sig(\mu)$. Then if $x \in [u = \lambda]$, we have $\eta_+(x) \geq \mu$.*

Proof. Let $X = \mathrm{cc}([u = \lambda], x)$ and $C = \mathrm{cc}([u < \mu], x)$. Assume first that X is a regional maximum. Then $X \in \mathcal{CC}([u \geq \lambda])$ and thus $\{X\} \in sig(\lambda)$. Therefore $\{X\} \in sig(\mu)$ and $\{X\} = sig(C)$. Since C contains a regional minimum, we conclude that X is both regional minimum and maximum, and therefore that $X = \overline{\Omega}$, which contradicts $\lambda < \max u$. Therefore X is not a regional maximum and X has at least one positive hole.

Let H be a positive hole of X. For some $\varepsilon > 0$, we have $H_\varepsilon \subseteq [u > \lambda]$ and let $\tilde{H} = cc([u > \lambda], H_{\underline{\varepsilon}})$. Let us assume that $\tilde{H} \subseteq [u < \mu]$. Let E be a regional maximum in $\tilde{H}$ with $u(E) \subset (\lambda, \mu)$; then $E \in sig(C)$, since $X \cup \tilde{H}$ is connected and contained in $[u < \mu]$, thus in C. If we suppose that $C \cap [u < \lambda] \neq \emptyset$, we have a regional minimum E' at level $< \lambda$ in C. Thus $\{E, E'\} \subseteq sig(C) \in sig(\mu)$, but E and E' are in different elements of $sig(\lambda)$ since E is at level $> \lambda$ and E' at level $< \lambda$. This contradiction shows that $C \subseteq [\lambda \leq u < \mu]$. Let $y \in \partial C$ (possible since $C \neq \overline{\Omega}$) and $D = \mathrm{cc}([u \geq \mu], y)$. Then D contains a regional maximum E' at level $\geq \mu$.

Since $C \cup D$ is connected and in $[u \geq \lambda]$, we deduce that E and E' are in the same element of $sig(\lambda)$. But E' and E cannot be in the same element of $sig(\mu)$ since the first is at level at least μ and the second at level $< \mu$. This contradiction proves that $\tilde{H} \cap [u \geq \mu] \neq \emptyset$ and there is a regional maximum $E_{\tilde{H}}$ in $\tilde{H}$ at level at least μ.

If there were two positive holes H and H' of X, and defining $\tilde{H}, \tilde{H}'$ as in last paragraph, we would have $\tilde{H} \cup \tilde{H}' \cup X \subseteq [u \geq \lambda]$, while it is a connected set. Thus $E_{\tilde{H}}$ and $E_{\tilde{H}'}$ would be in the same element of $sig(\lambda) = sig(\mu)$, and since they are in $[u \geq \mu]$, we would have a connected component of $[u \geq \mu]$ meeting H and H', and therefore X, which is impossible since $X \subseteq [u = \lambda]$. Therefore there is a unique positive hole H of X.

We can assume that ε is such that $H_\varepsilon \subseteq [\lambda < u < \mu]$. Let $y \in [u = \mu] \cap \partial\, \mathrm{cc}([\lambda < u < \mu], H_\varepsilon)$ and $Y = \mathrm{cc}([u = \mu], y)$. It is easy to see that $(X, Y) = \mathrm{cc}([\lambda < u < \mu], H_\varepsilon)$. If we assume that (X, Y) contains a regional extremum E, we can consider $\mathrm{cc}([u \geq \mu], Y)$, which has a regional maximum E', and then E and E' are in different elements of $sig(\mu)$. But $(X, Y) \cup \mathrm{cc}([u \geq \mu], Y)$ is connected and in $[u \geq \lambda]$, proving that E and E' are in the same element of $sig(\lambda)$, which is a contradiction with $sig(\lambda) = sig(\mu)$. Thus (X, Y) has no regional extremum and the holes of X not containing Y are all negative. By applying Proposition 4.20, we conclude that $[X, Y)$ is a monotone section, thus $\eta_+(x) \geq \lambda$. □

Proposition 4.33. *Let $u \in C(\overline{\Omega})$ be a weakly oscillating function. Let $\lambda \in \mathbb{R}$. If λ is a singular value of u, then λ is a critical value of u.*

Proof. Suppose that λ is a singular value which corresponds to a maximum value. Then there is a connected component X of $[u \geq \lambda]$ which does not intersect any connected component of $[u \geq \mu]$ for all $\mu > \lambda$. Then $X \in sig([u \geq \lambda])$ and $X \notin sig([u \geq \mu])$ for any $\mu > \lambda$. Thus λ is a critical value of u.

If λ is a minimum value, there is $q \in [u = \lambda]$ such that, if $\mu > \lambda$, then $\mathrm{cc}([u < \mu], q) \neq \emptyset$ and $\mathrm{cc}([u < \lambda], q) = \emptyset$. If $X = \mathrm{cc}([u = \lambda], q)$, then $X \in sig([u < \mu])$, $X \notin sig([u < \lambda])$. Thus λ is a critical value of u.

Thus, we may assume that there are no regional extrema at level λ and either $\eta_+(x) = \lambda$, or $\eta_-(x) = \lambda$. In case $\eta_+(x) = \lambda$, we deduce that λ is a critical value of u, by an application of Proposition 4.32. If $\eta_-(x) = \lambda$ and λ is not a critical value of u, then $sig(\mu) = sig(\lambda)$ for all μ in a neighborhood of λ. Proceeding as in the proof of Proposition 4.32 we would obtain that $C = \mathrm{cc}([u = \lambda], x)$ can be extended to the left as a monotone section, i.e., that $\eta_-(x) < \lambda$, a contradiction. We conclude that λ is a critical value of u. □

Remark 4.34. From Remark 4.8 and the above results it follows that λ is a critical value of u if and only if $-\lambda$ is a critical value of $-u$.

Our purpose in next section will be to prove that the singularities of the tree of shapes of an image coincide with the notion of singular value defined here. Hence all notions of singularities defined in this chapter are equivalent.

4.5 Singular Values of the Tree Versus Singular Values

Throughout all this section we shall assume that p_∞ is a regional extremum of u.

Definition 4.35. Let $u \in \mathcal{USC}(\overline{\Omega})$. Let S be a limit shape of u. We say that S may be extended to the right (left) if there is a limit shape T of u with $S \subseteq T$ (resp., $T \subseteq S$) , $S \neq T$, such that $[S, T]$ (resp., $[T, S]$) is a monotone section.

Definition 4.36. Let $u \in \mathcal{USC}(\overline{\Omega})$, and S be a limit shape of u. We say that $S \neq \overline{\Omega}$ is singular shape if S is either a leaf, or S cannot be extended to the left or to the right. $S = \overline{\Omega}$ is a singular shape if it cannot be extended to the left.

Recall that by Proposition 4.26 limit shapes can be associated to a certain level λ. In case that we are considering the shape $S = \overline{\Omega}$ as a limit shape of upper (resp. lower) type we shall take the supremum (resp. the infimum) of the set of values λ for which $\overline{\Omega} = \mathrm{Sat}(cc([u \geq \lambda]))$ (resp. $\overline{\Omega} = \mathrm{Sat}(cc([u < \lambda])))$ Thus the following definition has sense.

Definition 4.37. Let $u \in C(\overline{\Omega})$. We say that λ is a singular value of the tree of shapes of u if there is a limit shape S corresponding to the level λ such that S is a singular shape.

We notice that if $u \in C(\overline{\Omega})$ is a weakly oscillating function, then the results of Sect. 2.5 hold. That is, there is a finite number of leaves and a finite number of maximal monotone sections of the tree of shapes of u. Hence, there is a finite number of singularities of the tree.

Proposition 4.38. *Let $u \in C(\overline{\Omega})$ and $\lambda \in \mathbb{R}$. Then λ is a singular value of the tree of shapes of u if and only if $-\lambda$ is a singular value of the tree of shapes of $-u$.*

Proof. Since the limit shapes of u corresponding to the level λ coincide with the limit shapes of $-u$ corresponding to the level $-\lambda$, the statement follows easily from the definition of singular values of the tree. □

Lemma 4.39. *Leaves are shapes, or limit shapes, which are regional extrema. In particular, leaves are singular shapes.*

Proof. Assume that M is a regional maximum at height λ which is a shape. Then for any $\epsilon > 0$ the set $\mathrm{cc}([\lambda - \epsilon < u \leq \lambda], M)$ is a neighborhood of M and therefore $\mathrm{cc}([u \geq \lambda], M) = M$. Let S be a shape contained in M. Thus $S \subseteq [u = \lambda]$. If $S = \mathrm{Sat}(\mathrm{cc}([u < \mu], p))$ for some $\mu \in \mathbb{R}$, $p \in S$, then $\lambda < \mu$. Since $[u < \mu]$ is open in $\overline{\Omega}$, then S contains a neighborhood of M, a contradiction. Thus $S = \mathrm{Sat}(\mathrm{cc}([u \geq \mu], p))$. Since $u = \lambda$ on S, we have that $\lambda \geq \mu$. If $\mu < \lambda$, then $[u \geq \lambda] \subseteq [u > \mu] \subseteq [u \geq \mu]$ and, since $[u > \mu]$ is open, then $S = \mathrm{cc}([u \geq \mu], p)$ is a neighborhood of M, a contradiction. Thus, we have $\mu = \lambda$, and $S = \mathrm{cc}([u \geq \lambda], p) = M$. We conclude that S is a leaf.

If M is a regional minimum which is a limit shape of the tree of shapes of u, then M is a regional maximum which is a shape of the tree of shapes of $-u$. Then M is a leaf of the tree of shapes of $-u$, hence also of u.

If M is a leaf, then it is a limit shape. Hence it is a connected component of an upper $[u \geq \lambda]$ or lower level set $[u \leq \lambda]$, $\lambda \in \mathbb{R}$. It is easy to check that it is a connected component of $[u = \lambda]$. Then it is a regional extremum. □

Theorem 4.40. *Let $u \in C(\overline{\Omega})$ be a weakly oscillating function. Let λ be a singular value of u. Then λ is a singular value of the tree of shapes of u.*

Proof. Let λ be a singular value of u. Then there is a point $x \in \overline{\Omega}$ such that either $\eta_+(x) = \lambda$, or $\eta_-(x) = \lambda$. Let $X = \mathrm{cc}([u = \lambda], x)$. If $p_\infty \in X$, then X is a regional extremum. Assume that X is a regional maximum. Observe that $S_\lambda^+ = \mathrm{Sat}(\mathrm{cc}([u \geq \lambda], X)) = \overline{\Omega}$. Observe that if T is any shape of upper type $T \neq S_\lambda^+$, then T is contained in the saturation of a connected component of $[u < \lambda]$, thus in a shape S of lower type. In other words, S_λ^+ cannot be extended to the left as a monotone section of the tree: λ is a singular value of the tree of shapes of u.

Assume that X is a regional minimum. Then $S_\lambda^- = \mathrm{Sat}(\mathrm{cc}([u \leq \lambda], X)) = \overline{\Omega}$. If T is any shape of lower type $T \neq S_\lambda^-$, then T is contained in the saturation of a connected component of $[u \geq \mu]$, for some $\mu > \lambda$, thus in a shape S of upper type. In other words, S_λ^- cannot be extended to the left as a monotone section of the tree: λ is a singular value of the tree of shapes of u.

Thus, we may assume that $p_\infty \notin X$. Observe that, by Remark 4.8 and Proposition 4.38, it is sufficient to consider the case $\eta_+(x) = \lambda$. In that case, by Proposition 4.22, we know that $X = \mathrm{cc}([u = \lambda], x)$ has, at least, two positive holes. If all its holes are positive, then X is a regional minimum and $S = \mathrm{Sat}(\mathrm{cc}([u \leq \lambda], X)$ is a singular shape since it cannot be extended to the left as a monotone section of the tree. Thus we may assume that X has also a negative hole.

If the external hole is of positive type, then

$$\mathrm{Sat}(X) = \mathrm{Sat}(\mathrm{cc}([u \leq \lambda], X))$$

and is a limit shape (of lower type). If the external hole is of negative type, then

$$\mathrm{Sat}(X) = \mathrm{Sat}(\mathrm{cc}([u \geq \lambda], X))$$

and is a shape of upper type. Let T be a shape contained in $\mathrm{Sat}(X)$, $T \neq \mathrm{Sat}(X)$. Observe that $T \cap X = \emptyset$, otherwise $X \subseteq T$, and, thus, $\mathrm{Sat}(X) \subseteq T$, a contradiction. Thus T is contained in an internal hole of X, say H_1. We know that there is a second internal hole H_2 of X. Let $p \in H_2$. We may choose such a point so that $u(p) \neq \lambda$. If $u(p) > \lambda$ there is a shape of upper type T' contained in H_2, hence disjoint to T. If $u(p) < \lambda$ there is a shape of lower type T' contained in H_2, hence disjoint to T. Thus $\text{Sat}(X)$ cannot be extended to the left. We conclude that λ is a singular value of the tree of shapes of u. □

Theorem 4.41. *Let $u \in C(\overline{\Omega})$ be a weakly oscillating function. Let λ be a singular value of the tree of shapes of u. Then λ is a singular value of u.*

Proof. Suppose that λ is not a singular value of u. In particular, λ is not the value of a regional extremum. Then for any $x \in [u = \lambda]$ we have that $\eta_-(x) < \lambda < \eta_+(x)$. Let us fix such an x. We know that $X = \text{cc}([\eta_-(x) < u < \eta_+(x)], x)$ is a monotone section. We know that for each $\mu \in (\eta_-(x), \eta_+(x))$ the set $X_\mu = [u = \mu] \cap X$ is connected, and has a single positive and a single negative hole. Moreover these sets are nested. Two cases are to be considered: a) for any $\mu_1 < \mu_2$ in $(\eta_-(x), \eta_+(x))$, the set X_{μ_1} is contained in an internal hole of X_{μ_2}, b) for any $\mu_1 < \mu_2$ in $(\eta_-(x), \eta_+(x))$, the set X_{μ_2} is contained in an internal hole of X_{μ_1}.

Suppose that we are in case a). Let $\eta_-(x) < \alpha' < \alpha < \mu < \beta < \eta_+(x)$. Let $X_\mu^- = \text{cc}([u < \mu], X_{\alpha'})$, $S_\mu^- = \text{Sat}(X_\mu^-)$. We claim that the interval of shapes $[S_\alpha^-, S_\beta^-]$ is a monotone section of the tree. First, observe that $S_\beta^- \setminus S_\alpha^- \subseteq X \cap [\alpha \leq u < \beta]$. If $T \in [S_\alpha^-, S_\beta^-]$ is a shape of upper type such that $T \neq S_\alpha^-, S_\beta^-$, then we may write $T = \text{Sat}(Z)$ where $Z \in \mathcal{CC}([u \geq \overline{\mu}])$. Since $\partial T \subseteq \partial Z \subseteq X_{\overline{\mu}}$, then $\overline{\mu} \in [\alpha, \beta)$. Then $Z \supseteq X_\mu$ for any $\mu \in [\overline{\mu}, \eta_+(x))$. Then $T = \text{Sat}(Z) \supseteq \text{Sat}(X_\mu)$ for any $\mu \in [\overline{\mu}, \eta_+(x))$. This implies that $T \supseteq S_\beta^-$, a contradiction. If T is a shape contained in S_β^- and disjoint to S_α^-, then $T \subseteq X \cap [\alpha \leq u < \beta]$. This implies that $S_\alpha^- \subseteq T$, a contradiction. We have proved our claim.

Suppose that we are in case b), and let $\eta_-(x) < \alpha < \mu < \beta < \eta_+(x)$. Let $X_\mu^+ = \text{cc}([u \geq \mu], X_{\alpha'})$, $S_\mu = \text{Sat}(X_\mu^+)$. As in the last paragraph, we prove that the interval of shapes $[S_\beta^+, S_\alpha^+]$ is a monotone section of the tree.

Let $S \neq \overline{\Omega}$ be a limit shape of u corresponding to the level λ and let $x \in \partial S \subseteq [u = \lambda]$. We use the notation of the first paragraph of the proof. We know that $\eta_-(x) < \lambda < \eta_+(x)$ and $X = \text{cc}([\eta_-(x) < u < \eta_+(x)], x)$ is a monotone section. We know that S has one of the forms described in Proposition 4.26, and by Lemma 2.10 of Chap. 2, we have that $S \subseteq \text{Sat}(\partial S) \subseteq \text{Sat}(X_\lambda)$. On the other hand, we observe that if we are in case a), then $\text{Sat}(X_\lambda) \cap [u < \lambda] \subseteq S \cap [u < \lambda]$, and $\text{Sat}(X_\lambda) \cap [u > \lambda] \subseteq S \cap [u > \lambda]$ if we are in case b). Thus, by the previous part of the proof we know that $\text{Sat}(X_\lambda)$ and S are contained in a monotone section of the tree. Hence S is not a singular shape. We conclude that λ cannot be a singular value of the tree of shapes of u. □

4.6 Review on the Topographic Description of Images

The use of a topographic description of images, surfaces, or $3D$ data has been introduced and motivated in different areas of research, including image processing, computer graphics, and geographic information systems (GIS), e.g., [8–11, 14, 18, 23, 34, 40, 52, 57, 78, 79, 92, 98, 107, 113]. The motivations for such a description differ depending on the field of application. In all cases these descriptions aim to achieve an efficient description of the basic shapes in the given image and their topological changes as a function of a physical quantity that depends on the type of data (height in data elevation models, intensity in images, etc.). In our brief literature review we have separated the works into two main areas of research: computer graphics and image processing. In some cases this separation is somewhat arbitrary, some papers, if not all, could be included in both areas, since the application could be oriented to one or the other.

In computer graphics and geographic information systems, topographic maps represent a high level description of the data. Topographic maps are represented by the contour maps, i.e., the isocontours of the given scalar data. The description of the varying isocontours requires the introduction of data structures, like the *topographic change tree* or *contour tree* which can represent the nesting of contour lines on a contour map (or a continuous topographic structure) [52, 90, 113]. In all cases, the proposed description can be considered as an implementation of Morse theory, in the sense that Morse theory describes the topological change of the isocontours of scalar data or height function as the height varies, and relates these topological changes to the criticalities of the function. Given the scalar data u defined in a domain Ω of $\mathbb{R}^N$ ($u : \Omega \to \mathbb{R}$), the contour map is defined in the literature as the family of isocontours $[u = \lambda]$, $\lambda \in \mathbb{R}$, or in terms of the boundaries of upper (or lower) level sets $[u \geq \lambda]$ ($[u \leq \lambda]$). The first description is more adapted to the case of smooth data while the second description can be adapted to more general continuous data where there are plateaus of constant elevation or discontinuous data. The second description has been addressed in [23, 52], while the first description has been used in [8, 9, 113], where an *a priori* interpolation of the discrete data is required so that the regularity assumptions permit the isocontour description.

The contour map is organized in a data structure, either the contour tree [52, 113], or the Reeb graph [87, 110]. The contour tree represents the nesting of contour lines of the contour map. According to [52], each node represents a connected component of an upper (or lower) level set $[u \geq \lambda]$ ($[u \leq \lambda]$), and links between nodes represent a parent-child relationship, a link going from the containing to the contained set in the upper tree, or viceversa if we consider the lower tree. Each node has a list of descendants, its corresponding elevation value, a list of boundary points, and its parent. The contour tree encodes the topological changes of the level curves of the data. Critical values and its associated features, peaks (maxima), pits (minima), or passes (saddles), can be extracted from the contour trees [52]. The description of the topographic changes requires the use of both upper and lower trees, and the contour tree can also be used as a tool to compute other terrain features such as ridges and ravines [52]. For practical applications, the data structure has to be implemented with a fast algorithm and with minimal storage requirements. In [113] this is accomplished with a variant of the contour tree where the criticalities (maxima, minima, saddles, computed in a local way) are computed first. In [9] several attributes have been added to the contour data which can be used to select a subsampled family of contours which are representative of the data. As examples of such attributes the authors choose the length or area of the isocontours, the ratio length of the isocontour/area of the enclosed set, or the integral of the gradient along the isocontour.

A related data structure is the *Reeb graph*, which represents the splitting and merging of the isocontours. The *Reeb graph* of the height function u is obtained by identifying two points $p, q \in \Omega$ such that $u(p) = u(q)$ if they are in the same connected component of the isocontour $[u = u(p)]$. Thus, a cross-sectional contour corresponds to a point of an edge of the Reeb graph, and a vertex represents a critical point of the height function u. The Reeb graph was proposed in [110] as a data structure for encoding topographic maps. The authors proposed to compute it following the computation of the so-called surface network. The surface network is also a topological graph, i.e., a graph that represents the relations among critical points, whose vertex are critical points and the edges represent either a ridge or a ravine line. A ridge (ravine) line is a line with steepest gradient which joins a pass to a peak (pit) [47, 110]. The critical points are computed with an algorithm based on local computations, so that the Euler's formula relating the number of peaks, pits and passes, holds. Then ridges and ravines are computed following the steepest lines. In the context of computer graphics, Morse theory has also been used to encode surfaces in $3D$ space [107]. In [107], the authors also use a tree structure like the Reeb graph complemented with information about the Morse indexes of the singularities and including enough (information about) intermediate contours to be able to reconstruct by interpolation the precise way in which the surface is embedded in $3D$ space.

In image processing, the topographic description was advocated as a local and contrast invariant description of images (i.e., invariant under illumination changes), and has led to an underlying notion of shapes of an image as the family of connected components of upper or lower level sets of the image [18, 19, 92, 98]. An efficient description of the family of shapes in terms of a tree was proposed in [78, 79] and further developed in [57]. The tree of shapes as proposed in [78, 79] fuses the information of both the trees of upper $[u \geq \lambda]$ and lower $[u \leq \lambda]$ level sets of the scalar image u. As we have seen in Chap. 2 the key idea for this fusion is the notion of shape as a connected component of an upper $[u \geq \lambda]$ or lower $[u < \lambda]$ level set in which the holes are filled-in. This topographic structure has been further studied in [11, 78], where a weak Morse description of this topographic structure was developed.

The mathematical description permitted to include the case of images as upper semicontinuous functions. In [57], following bilinear interpolation of the discrete data, the image could be treated as a continuous function and a tree of bilinear level lines $[u = \lambda]$ was computed. The tree of bilinear level lines is more related to the contour tree computed with the isocontours of the interpolated image. A subtree containing the so-called meaningful level lines [25] can be extracted which contains the main level lines according to the distribution of gradient values of the image in a statistical way [25, 57]. The work in [50] can be considered as a mathematical description of the (iso) contour tree in the case of two-dimensional functions. In [23], Morse theory has also been used as a basic model to describe the geometric structures of $2D$ and $3D$ images, and in general, of multidimensional data. Applications have been given in different domains, in particular, to visualize structures in $3D$ medical images. The data are typically multi-dimensional sampled data, and it cannot be assumed that the function is Morse in a traditional sense, even if interpolated. Thus, the authors adapt Morse theory using combinatorial methods. The authors assume that the given data are interpolated by a continuous real valued function u. The basic geometric objects studied are the boundaries of the connected components of upper level sets $[u \geq \lambda]$ of u and their variation with the level λ. In their set of axioms, the authors assume that those boundaries are compact, oriented manifolds in $\mathbb{R}^N$, they precise their local structure, and its connection with the original sampled data. In particular, those axioms imply that the topological structure of the sampled data is reflected by any interpolating function satisfying their axioms. Then critical points and critical values are defined, obtaining maximum, minimum, and saddle critical points (and values). Criticalities are defined by local analysis of the function, including the case of degenerate sets (i.e., connected regions of the sampled data with the same values). The authors prove that the topology of their basic objects changes at a critical level, and does not change between critical levels. Then the criticality graph is defined, the vertices of the graph are the criticalities and the edges go between criticalities in such a way that no further criticalities are located between them. The regions represented by each edge are called zones of the critical point with higher value. In each zone the boundaries of the upper level sets are homeomorphic [23]. The topology change at a critical value is computed by combinatorial methods if the critical value is a saddle or a minimum, and by the genus in the case of a maximum value. In some sense, this structure is related to a Reeb graph. Efficient algorithms are proposed which compute the criticality graph [23].

Let us finally mention the use of topographic maps to encode digital elevation maps. In that case, the analysis of level lines permits to select those that reflect the topographic structure of the data and many interpolation algorithms have been designed in order to reconstruct it (see [21, 46, 108] and references therein).

Chapter 5
Merging the Component Trees

Chapter 2 started by presenting the tree of shapes as a fusion of the tree of connected components of upper level sets and the tree of connected components of lower level sets, known commonly as the component trees. We come back in this chapter to a constructive study of this fusion, leading to an algorithm applicable in any dimension. Direct and more efficient algorithms, specific to dimension 2, are presented in following chapters. However, the generality of the algorithm presented here and its natural implementation as the result of merging the component trees make it interesting.

5.1 The Structure of the Trees of Connected Components of Upper and Lower Level Sets

Recall that if $u : \overline{\Omega} \to \mathbb{R}$ is an upper semicontinuous function, $\mathcal{U}(u)$ and $\mathcal{L}(u)$ denote the trees of connected components of upper and lower level sets of u. To understand the structure of both trees, their leaves and maximal branches, it will be useful to describe their limit nodes.

Proposition 5.1. *(i) If X is a limit node of $\mathcal{U}(u)$, then either $X \in \mathcal{U}(u)$ or $X \in \mathcal{CC}([u > \lambda]$ for some $\lambda \in \mathbb{R}$.*
(ii) If X is a limit node of $\mathcal{L}(u)$, then either $X \in \mathcal{L}(u)$ or $X \in \mathcal{CC}([u \leq \lambda]$ for some $\lambda \in \mathbb{R}$.

Proof. Being identical, we just sketch the proof of (i). If X is a limit node of $\mathcal{U}(u)$, then X is an inf or a sup of an ordered set of upper connected components which we may assume countable. If $X = \cap_n X_n$ where $X_n \in \mathcal{CC}([u \geq \lambda_n])$ with $\lambda_n \uparrow \lambda$, then $X \in \mathcal{CC}([u \geq \lambda])$. If $X = \cup_n X_n$ where $X_n \in \mathcal{CC}([u \geq \lambda_n])$ with $\lambda_n \downarrow \lambda$, then $X \in \mathcal{CC}([u > \lambda])$. □

Proposition 5.2. *Let $u \in C(\overline{\Omega})$ be a weakly oscillating function. Then*

(i) If X is a leaf of $\mathcal{U}(u)$, then X is a regional maximum of u.
(ii) If X is a leaf of $\mathcal{L}(u)$, then X is a regional minimum of u.

V. Caselles and P. Monasse, *Geometric Description of Images as Topographic Maps*, Lecture Notes in Mathematics 1984,
DOI 10.1007/978-3-642-04611-7_5,

Proof. (i) By Proposition 5.1, if X is a leaf of $\mathcal{U}(u)$, then $X \in \mathcal{CC}([u \geq \lambda])$ for some $\lambda \in \mathbb{R}$. If $u(x) > \lambda$ for some $x \in X$, then the node $\mathrm{cc}([u \geq u(x)], x)$ is nonempty and contained in X. Thus $u = \lambda$ on X and $X \in \mathcal{CC}([u = \lambda])$. If $X = \overline{\Omega}$, our statement is obviously true. If $X \neq \overline{\Omega}$, then by Lemma 4.12 all holes of X must be of negative type. Hence $\mathrm{cc}([u > \lambda - \epsilon], X) = \mathrm{cc}([\lambda - \epsilon < u \leq \lambda], X)$, for any $\epsilon > 0$, and $\mathrm{cc}([u > \lambda - \epsilon], X)$ is an open set containing X.

Being similar to the proof of (i), we skip the proof of (ii). □

Proposition 5.3. *Assume that $u \in C(\overline{\Omega})$ is a weakly oscillating function. The tree $\mathcal{U}(u)$ has a finite number of leaves and a finite number of maximal branches. If $\mathcal{B} = [A, B]$ is a maximal branch of $\mathcal{U}(u)$ with A, B being limit nodes, then*
a) either $B = \overline{\Omega}$ or $B \in \mathcal{CC}([u > \lambda])$ for some $\lambda \in \mathbb{R}$. In the second case there is no bifurcation between A and B, and if we let $B' = \mathrm{cc}([u \geq \lambda], B)$, then $B' = \inf[B, \overline{\Omega}]$ and $[A, B']$ contains a bifurcation. Moreover, B cannot be a leaf unless u is constant.
b) $A \in \mathcal{CC}([u \geq \lambda])$ for some $\lambda \in \mathbb{R}$ and either A is a leaf, or for any $X \in \mathcal{U}(u)$, $X \subsetneq A$, $[X, A]$ contains a bifurcation.

Proof. Leaves of $\mathcal{U}(u)$ are regional maxima of u, hence there are finitely many of them. Let $\mathcal{B} = [A, B]$ be a maximal branch in $\mathcal{U}(u)$. By Proposition 5.1 either $B \in \mathcal{CC}([u \geq \lambda])$, or $B \in \mathcal{CC}([u > \lambda])$ for some $\lambda \in \mathbb{R}$. If $B \in \mathcal{CC}([u \geq \lambda])$ and $\lambda > \inf_{x \in \overline{\Omega}} u(x)$, then, using Lemma 4.28, we would be able to extend the branch $\mathcal{B}$ to the right. Hence, $\lambda = \inf_{x \in \overline{\Omega}} u(x)$, i.e. $B = \overline{\Omega}$. If $B \in \mathcal{CC}([u > \lambda])$, then $[A, B]$ does not contain a bifurcation. Let $B' = \mathrm{cc}([u \geq \lambda], B)$, then $[A, B']$ must contain a bifurcation, otherwise $\mathcal{B}$ would not be maximal. The argument in Lemma 4.28 proves that $B' = \inf[B, \overline{\Omega}]$. The last assertion follows from Proposition 5.2.

Now, observe that $A \in \mathcal{CC}([u \geq \lambda])$ for some $\lambda \in \mathbb{R}$. Since $\mathcal{B}$ is maximal, if A is not a leaf, then for any $X \in \mathcal{U}(u)$, $X \subsetneq A$, $[X, A]$ contains a bifurcation.

Since u has a finite number of regional maxima, there are finitely many maximal branches in $\mathcal{U}(u)$, since any two of them are disjoint. □

Proposition 5.4. *Assume that $u \in C(\overline{\Omega})$ is a weakly oscillating function. The tree $\mathcal{L}(u)$ has a finite number of leaves and a finite number of maximal branches. If $\mathcal{B} = [A, B]$ is a maximal branch of $\mathcal{L}(u)$ with A, B being limit nodes, then*
a) either $B = \overline{\Omega}$ or $B \in \mathcal{CC}([u < \lambda])$ for some $\lambda \in \mathbb{R}$. In the second case there is no bifurcation between A and B, and if we let $B' = \mathrm{cc}([u \leq \lambda], B)$, then $B' = \inf(B, \overline{\Omega}]$ and $[A, B']$ contains a bifurcation. Moreover, B cannot be a leaf unless u is constant.
b) $A \in \mathcal{CC}([u \leq \lambda])$ for some $\lambda \in \mathbb{R}$ and either A is a leaf, or for any $X \in \mathcal{L}(u)$, $X \subsetneq A$, $[X, A]$ contains a bifurcation.

We have used the notation $(B, \overline{\Omega}] = [B, \overline{\Omega}] \setminus \{B\}$.

Proof. Leaves are regional minima of u, hence there are finitely many of them. Let $\mathcal{B} = [A, B]$ be a maximal branch in $\mathcal{L}(u)$. By Proposition 5.1 either $B \in \mathcal{CC}([u \leq \lambda])$ or $B \in \mathcal{CC}([u < \lambda])$ for some $\lambda \in \mathbb{R}$. If $B \in \mathcal{CC}([u \leq \lambda])$ and $\lambda < \sup_{x \in \overline{\Omega}} u(x)$, then, using the arguments in Lemma 4.28, we would be able to extend the branch $\mathcal{B}$ to the right. Hence, $\lambda = \sup_{x \in \overline{\Omega}} u(x)$ and $B = \overline{\Omega}$.

If $B \in \mathcal{CC}([u < \lambda])$, then $[A, B]$ does not contain a bifurcation. Let us prove that $[A, B']$ contains a bifurcation, where $B' = \mathrm{cc}([u \leq \lambda], B)$. If $B' = \overline{\Omega}$ and it does not contain a bifurcation we are in the previous case for B (we could take $B = B'$). Thus, we may assume either that $B' = \overline{\Omega}$ and $[A, B']$ contains a bifurcation, or $B' \neq \overline{\Omega}$, that is $\lambda < \sup_{x \in \overline{\Omega}} u(x)$. We have to consider only the last case. By (the proof of) Lemma 4.28, there is an $\epsilon > 0$ such that if $\mu \in (\lambda, \lambda + \epsilon)$, then $C := \mathrm{cc}([u < \mu], B')$ does not contain any other connected component of $[u \leq \lambda]$ besides B' and contains the same regional extrema as B'. If there is no bifurcation in $[B, C]$, then we can extend $[A, B]$ to the right. Thus, we may assume that $[B, C]$ contains a bifurcation, i.e., there is $Y \in \mathcal{CC}([u < \alpha])$ with $Y \cap B = \emptyset$ and $Y \subseteq C$. Notice that we have $\alpha \leq \mu$. Let us prove that we may assume that $\alpha \leq \lambda$. If $\lambda < \alpha \leq \mu$ and $Y \cap [u \leq \lambda] \neq \emptyset$, then we take V to be a connected component of $[u \leq \lambda]$ inside Y. Then C contains V and B', but this is not possible in view of our choice of C. Hence $Y \subseteq [\lambda < u \leq \mu]$, and we deduce that Y contains a regional minimum of u not in B', hence also C does it, a contradiction with our choice of C. Thus, we may assume that $\alpha \leq \lambda$. Let $V := \mathrm{cc}([u < \lambda], Y)$. Since $Y \cap B = \emptyset$, we have that $V \cap B = \emptyset$. If $\mathrm{cc}([u \leq \lambda], V)$ is disjoint to B', then C contains two connected components of $[u \leq \lambda]$, again a contradiction with our choice of C. Hence, $B' = \mathrm{cc}([u \leq \lambda], V)$, and in this case $[A, B']$ contains a bifurcation since B and V are disjoint. Finally, the argument in Lemma 4.28 proves that $B' = \inf(B, \overline{\Omega}]$. The last assertion in a) follows from Proposition 5.2.

Since A is a limit node, then $A \in \mathcal{CC}([u \leq \lambda])$ for some $\lambda \in \mathbb{R}$. Let us prove that if A is not a leaf, then for any $X \in \mathcal{L}(u)$, $X \subsetneq A$, $[X, A]$ contains a bifurcation, i.e., there is $Y \in \mathcal{L}(u)$, $Y \subseteq A$ and $Y \cap X = \emptyset$. Fix such an X. Observe that $\inf_A u < \lambda$, otherwise $A \in \mathcal{CC}([u = \lambda])$ and A would be a leaf. If there are more than one connected components of $[u < \lambda]$ in A, our assertion is true. Thus, we may assume that there is only one connected component of $[u < \lambda]$ in A. Let $S \in [A, B]$, $S \in \mathcal{CC}([u < \lambda'])$ with $\lambda < \lambda'$. Observe that $[X, S]$ contains a bifurcation otherwise we could enlarge $[A, B]$ to the left. Let $Y \in \mathcal{L}(u)$ be such that $Y \subseteq S$ and $Y \cap X = \emptyset$. Notice that we may write that $X \in \mathcal{CC}([u < \alpha])$ with $\alpha \leq \lambda$ and $Y \in \mathcal{CC}([u < \mu])$ for some $\mu < \lambda'$ (if $\mu = \lambda'$, then $Y = S$, contradicting the fact that $X \cap Y = \emptyset$). Since there is only one connected component of $[u < \lambda]$ in S (otherwise there would be a bifurcation in $[A, S]$ since A identifies one of them), we have that either $Y \subseteq A$ or $Y \subseteq [\lambda \leq u < \mu]$. In the first case, we have that there is a bifurcation in $[X, A]$. Let us consider the second case: if $Y \cap A = \emptyset$, then $[A, B]$ contains a bifurcation. Hence $Y \cap A \neq \emptyset$. Since Y is a node and A a limit node, then either $A \subseteq Y$, or $Y \subseteq A$. In the first case, we obtain a contradiction since

$X \subseteq A$ and $X \cap Y = \emptyset$. In the second case we have $Y \in \mathcal{CC}([u = \lambda])$, $Y = A$, and A would be a leaf. We conclude that there are more than one connected components of $[u < \lambda]$ in A and $[X, A]$ contains a bifurcation.

Since u has a finite number of regional maxima, there are finitely many maximal branches in $\mathcal{L}(u)$, since any two of them are disjoint. □

5.2 Construction of the Tree of Shapes by Fusion of Upper and Lower Trees

Let us recall the following result which is essentially a rephrasing of Lemma 2.12.

Lemma 5.5. *Let $u \in C(\overline{\Omega})$ be a weakly oscillating function, $\lambda \in \mathbb{R}$. The family of internal holes of all connected components of $[u \geq \lambda]$ coincides with* $\{\mathrm{Sat}(O) : O \in \mathrm{cc}([u < \lambda])\}$.

Proof. By Lemma 2.12, if $X \in \mathcal{CC}([u \geq \lambda])$ and Y is an internal hole of X, then there is $O \in \mathcal{CC}([u < \lambda])$ such that $Y = \mathrm{Sat}(O)$. Conversely, let $O \in \mathcal{CC}([u < \lambda])$, and observe that $\overline{\Omega} \setminus \mathrm{Sat}(O)$ is the external hole of O. By taking a point q in $\mathrm{Sat}(O)$ as the point at infinity, we may adapt the proof of Lemma 2.12 to obtain that there exists $Z \in \mathcal{CC}([u \geq \lambda])$ such that $\overline{\Omega} \setminus \mathrm{Sat}(O) = \mathrm{Sat}(Z, q)$. Then $\mathrm{Sat}(O)$ is a hole of Z. □

Since the shapes of the tree $\mathcal{S}(u)$ are the saturations of the nodes of $\mathcal{U}(u)$ and $\mathcal{L}(u)$, we can construct $\mathcal{S}(u)$ by fusing the information of the upper and lower trees. This operation can de done very simply because of the precise branch structure of both trees described in Propositions 5.3 and 5.4.

We denote by $[A, B]_+$ the interval of the upper tree determined by nodes $A, B \in \mathcal{U}(u)$, and $[A, B]_-$ the interval of the lower tree determined by nodes $A, B \in \mathcal{L}(u)$.

To fix ideas, let us assume that p_∞ is a global minimum of u and let $\Lambda = \mathrm{cc}([u = \inf_{x \in \overline{\Omega}} u(x)], p_\infty)$. Observe that Λ is a leaf of $\mathcal{L}(u)$. Let $\mathcal{L}_\Lambda(u) = \mathcal{L}(u) \setminus [\Lambda, \overline{\Omega}]_-$. All nodes of $[\Lambda, \overline{\Omega}]_-$ have $\overline{\Omega}$ as saturation. If $C \in \mathcal{L}_\Lambda(u)$, then all nodes previous to it do not contain Λ. Thus, $\mathcal{L}_\Lambda(u)$ is a union of maximal branches of $\mathcal{L}(u)$. Notice that the only node of $\mathcal{U}(u)$ containing p_∞ is $\overline{\Omega}$.

The trees $\mathcal{U}(u)$ and $\mathcal{L}_\Lambda(u)$ are broken into maximal branches of the form $[A, B]_i$, $i \in \{+, -\}$, where A is either a leaf or a bifurcating node and B may be a bifurcating node or not, in which case it coincides with $\overline{\Omega}$. Each branch $\mathcal{B} = [A, B]_i$, $i \in \{+, -\}$, determines the set of shapes $\mathrm{Sat}(\mathcal{B}) := \{\mathrm{Sat}(C) : C \in \mathcal{B}\}$ of $\mathcal{S}(u)$. This will be proved in Lemma 5.6. Then, given $[A, B]_i$, it suffices to describe how to link the upper end of the interval to another interval either of $\mathcal{U}(u)$ or $\mathcal{L}_\Lambda(u)$.

Lemma 5.6. *Let $A, B \in \mathcal{U}(u)$ (resp. $\mathcal{L}(u)$) such that $[A, B]_+$ (resp. $[A, B]_-$) is a branch and $p_\infty \notin B$. Then* $\mathrm{Sat}([A, B]_+)$ *(resp.* $\mathrm{Sat}([A, B]_-)$*) is an interval of $\mathcal{S}(u)$.*

Proof. The proofs being similar, we show the result when $A, B \in \mathcal{U}(u)$. Obviously it is sufficient to prove it when $[A, B]_+$ is a maximal branch. According to Proposition 5.3, we can write $A \in \mathcal{CC}([u \geq \lambda])$ and $B \in \mathcal{CC}([u > \mu])$ with $\mu < \lambda$ as $A \subseteq B$.

If $C \in \mathcal{U}(u)$ and $C \notin [A, B]_+$, then clearly $\mathrm{Sat}(C) \notin [\mathrm{Sat}(A), \mathrm{Sat}(B)]$ (interval of $\mathcal{S}(u)$). We therefore assume $C \in \mathcal{CC}([u < \nu])$ and $\mathrm{Sat}(A) \subseteq \mathrm{Sat}(C) \subseteq \mathrm{Sat}(B)$, which will lead to a contradiction.

If $\nu \leq \mu$, then B is in a hole H of C. If H is an internal hole, we get $\mathrm{Sat}(B) \subseteq H \subseteq \mathrm{Sat}(C)$, a contradiction. If H is the external hole of C, we have $\mathrm{Sat}(C) \cap B = \emptyset$ whereas $A \subseteq \mathrm{Sat}(C)$ and $A \subseteq B$, a contradiction.

If $\mu < \nu$, observe first that $\partial\,\mathrm{Sat}(C) \subseteq B$. Indeed, $\partial\,\mathrm{Sat}(C) \subseteq [u = \nu]$ and is connected, so either is in B or in a hole of B. It cannot be in a hole H since we would have that $A \subseteq \mathrm{Sat}(C) \subseteq \mathrm{Sat}(\partial\,\mathrm{Sat}(C)) \subseteq H$, a contradiction with $A \subseteq B$. Let $D = \mathrm{cc}([u \geq \nu], \partial\,\mathrm{Sat}(C))$. Then D is in the external hole of C and $D \subseteq B$. Since $A \subseteq \mathrm{Sat}(C)$, we have that $A \cap D = \emptyset$. Hence, there is a bifurcation between A and B in $\mathcal{U}(u)$, which is contrary to the hypothesis that $[A, B]_+$ is a branch. □

In Fig. 5.1 we show an image and its trees $\mathcal{U}(u)$ (middle) and $\mathcal{L}(u)$ (right). At the left of each node we see a schematic representation of the level set. In Figs. 5.2 and 5.3 we display the fusion of both trees for two different choices of the point at infinity, the point p in the case shown in Fig. 5.2 and the point q in the case of Fig. 5.3. Notice that the point p is a global minimum of u, while the point q is not. The choice of the point at infinity is arbitrary and the structure of the tree does not depend on it, though its concrete presentation may look different (the root can be different). In both figures, the left image displays the decomposition of both trees into maximal branches and encircled in a pointed curve we display all nodes of the lower tree containing the point at infinity, hence having $\overline{\Omega}$ as saturation. Both trees are fused according to the rules described below. Notice that the structure of the upper tree is linear while the lower one is not. Thus, the fusion of the trees illustrates the case of fusion of the maximal branches of $\mathcal{L}(u)$. Since they are similar to the fusion rules for the upper tree, the figures describe the generic situation.

Before explaining precisely the fusion rules for maximal branches, let us expose their principle. If $[A, B]_\pm$ is a maximal branch of $\mathcal{U}(u)$ or $\mathcal{L}(u)$, B is an open set defined by a strict inequality at a level λ. We can then consider the exterior boundary of B, $\partial\,\mathrm{Sat}(B)$ and the two sets $\mathrm{cc}([u \geq \lambda], \partial\,\mathrm{Sat}(B))$ and $\mathrm{cc}([u \leq \lambda], \partial\,\mathrm{Sat}(B))$ (one is designed by B' and the other by N below). They are limit nodes in $\mathcal{U}(u)$ and $\mathcal{L}(u)$ respectively, and their saturations are limit nodes in $\mathcal{S}(u)$. These saturations are therefore nested and we link $\mathrm{Sat}(B)$ to the smaller one, call it $\mathrm{Sat}(C)$. Then the interval $[\mathrm{Sat}(B), \mathrm{Sat}(C)]$ of $\mathcal{S}(u)$ has no other element than $\mathrm{Sat}(B)$ and $\mathrm{Sat}(C)$ themselves.

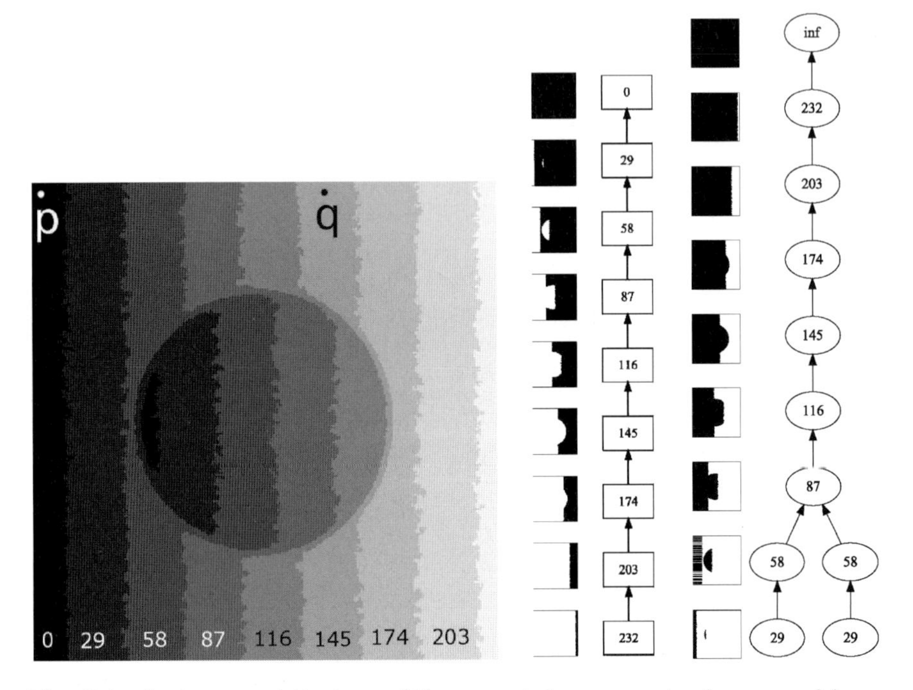

Fig. 5.1 An image and the trees of its connected components of upper and lower level sets (component trees). Left: original image. The numbers at the bottom of the image denote the gray levels. The points p and q are two possible points of infinity. Middle: the tree of connected components of upper level sets $\mathcal{U}(u)$. They are denote as squares with the value of their defining gray level inside. Right: the tree of connected components of lower level sets $\mathcal{L}(u)$. They are denote as ellipses. In both trees, the level sets at depicted in black at the left side of the nodes.

From now on, when using limit nodes we shall consider them as parts of the corresponding tree.

Fusion rules for maximal branches of $\mathcal{U}(u)$. Let $[A, B]_+$ be a maximal branch of $\mathcal{U}(u)$. By Proposition 5.3, either $B = \overline{\Omega}$ or $B \in \mathcal{CC}([u > \lambda])$ for some $\lambda \in \mathbb{R}$. In the first case, there is nothing to say. In the second case there is no bifurcation between A and B, and if we let $B' = \mathrm{cc}([u \geq \lambda], B)$, then $[A, B']_+$ contains a bifurcation. Observe that, by applying Lemma 5.5 to $-u$, we have that $\mathrm{Sat}(B)$ is a hole of a connected component N on $[u \leq \lambda]$. We compare the two limit shapes $\mathrm{Sat}(B')$ and $\mathrm{Sat}(N)$ (they are limit shapes by Proposition 4.26). Since they are limit shapes of different type, either they are different or both coincide with $\overline{\Omega}$. If both coincide with $\overline{\Omega}$, then $A \subseteq N$, and we may link $\mathrm{Sat}(B)$ to $\overline{\Omega}$. Thus we may assume that $\mathrm{Sat}(B') \neq \mathrm{Sat}(N)$. Since $\mathrm{Sat}(N)$ intersects $\partial B \subseteq B'$, then either $\mathrm{Sat}(N) \subsetneq \mathrm{Sat}(B')$ or $\mathrm{Sat}(B') \subsetneq \mathrm{Sat}(N)$. In the first case, we link $\mathrm{Sat}(B)$ to $\mathrm{Sat}(N)$ which is the limit shape generated by N. Notice that N is the lower end of a (not necessarily maximal)

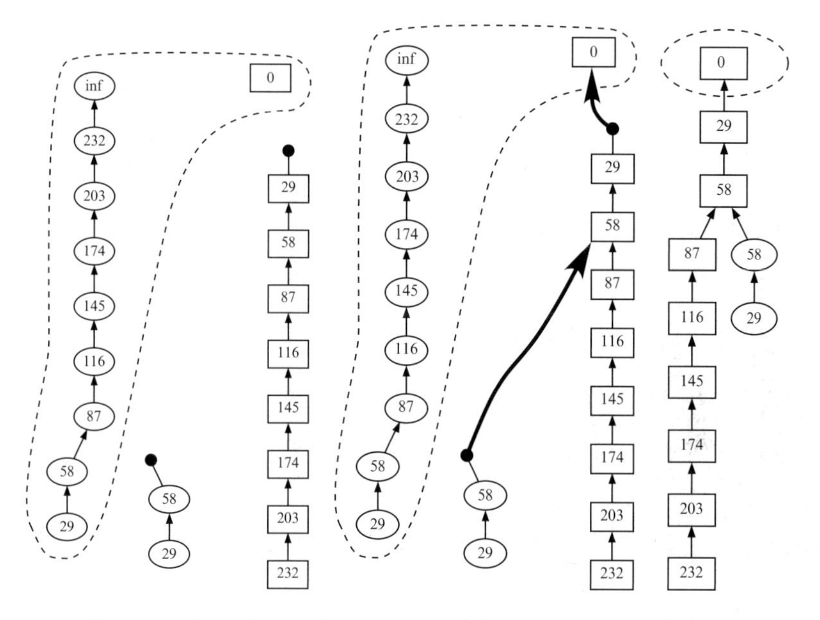

Fig. 5.2 Example of the fusion of both trees $\mathcal{U}(u)$ and $\mathcal{L}(u)$ for the image in Fig. 5.1. The point p (at level 0) is taken as the point at infinity. Left: the lower level sets containing p are surrounded by a pointed curve. Their saturation gives the root. We used the value 0 to identify the root. The rest of the trees is decomposed in maximal branches. The nodes that must be connected to the other tree are shown. Middle: the arrows show the connections between the maximal branches of both trees. Right: the tree of shapes obtained by fusion of $\mathcal{U}(u)$ and $\mathcal{L}(u)$. The root is surrounded by a pointed line. In this case, squares and ellipses denote shapes of upper and lower type, respectively.

branch of $\mathcal{L}(u)$. In the second case, we link $\mathrm{Sat}(B)$ to $\mathrm{Sat}(B')$. In this case B' is the lower end of a maximal branch of $\mathcal{U}(u)$.

Fusion rules for maximal branches of $\mathcal{L}(u)$. Let $[A, B]_-$ be a maximal branch of $\mathcal{L}(u)$. By Proposition 5.4, either $B = \mathrm{cc}([u \leq \lambda]) = \overline{\Omega}$ or $B \in \mathcal{CC}([u < \lambda])$ for some $\lambda \in \mathbb{R}$. In the first case, there is nothing to say. In the second case there is no bifurcation between A and B, and if we let $B' = \mathrm{cc}([u \leq \lambda], B)$, then $[A, B']_-$ contains a bifurcation. Observe that, by Lemma 5.5, we have that $\mathrm{Sat}(B)$ is a hole of a connected component N on $[u \geq \lambda]$. Again,we compare the two limit shapes $\mathrm{Sat}(B')$ and $\mathrm{Sat}(N)$. Since they are limit shapes of different type, either they are different or both coincide with $\overline{\Omega}$. If both coincide with $\overline{\Omega}$, then $\mathrm{Sat}(N) = \overline{\Omega}$ and $\lambda = \inf_{x \in \overline{\Omega}} u(x)$ but this is not the case, due to our choice of B. Thus we may assume that $\mathrm{Sat}(B') \neq \mathrm{Sat}(N)$. Since $\mathrm{Sat}(N)$ intersects $\partial B \subseteq B'$, then either $\mathrm{Sat}(N) \subsetneq \mathrm{Sat}(B')$ or $\mathrm{Sat}(B') \subsetneq \mathrm{Sat}(N)$. In the first case, we link $\mathrm{Sat}(B)$ to $\mathrm{Sat}(N)$ which is the shape generated by N. Notice that N is the lower end of a (not necessarily maximal) branch of $\mathcal{U}(u)$. In the second case, we link $\mathrm{Sat}(B)$ to $\mathrm{Sat}(B')$. In this case B' is the lower end of a maximal branch of $\mathcal{L}(u)$.

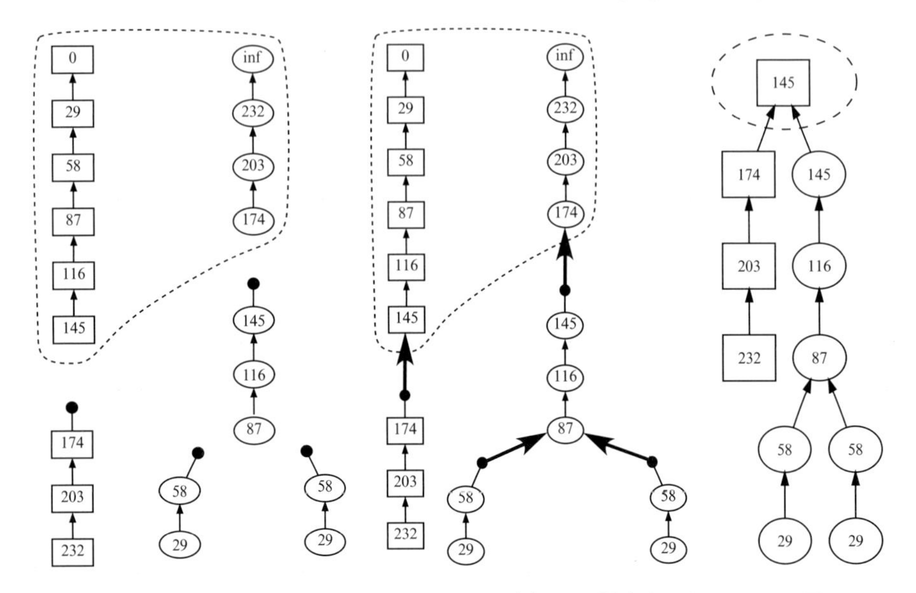

Fig. 5.3 Example of the fusion of both trees $\mathcal{U}(u)$ and $\mathcal{L}(u)$ for the image in Fig. 5.1. This time the point q (at level 145) is taken as the point at infinity. Left: the lower level sets containing q are surrounded by a pointed curve. Their saturation gives the root. We used the value 145 to identify the root. The rest of the trees is decomposed in maximal branches. The nodes that must be connected to the other tree are shown. Middle: the arrows show the connections between the maximal branches of both trees. Right: the tree of shapes obtained by fusion of $\mathcal{U}(u)$ and $\mathcal{L}(u)$. The root is surrounded by a pointed line. Again, in this case, squares and ellipses denote shapes of upper and lower type, respectively. Observe that the tree of shapes obtained here is equivalent to the one displayed in Fig. 5.2.

Observe that:

(i) the final structure contains the saturations of all connected components of upper and lower level sets,

(ii) If A is a node (or limit node) in $\mathcal{L}(u)$ which is the upper end of an interval, and B a node (or limit node) in $\mathcal{U}(u)$ which is the lower end of an interval, or viceversa, and we have linked $\mathrm{Sat}(A)$ to $\mathrm{Sat}(B)$, we have chosen $\mathrm{Sat}(B)$ to be the minimal limit shape in $\mathcal{U}(u)$ containing A. With this rule, we have not created cycles in the fusion of both trees. Thus, we have constructed a tree.

In Figs. 5.2 and 5.3 we display the decomposition of both trees $\mathcal{U}(u)$ and $\mathcal{L}(u)$ into maximal branches and their fusion. First the nodes whose saturation is $\overline{\Omega}$ have been separated and represented by a single shape, the root of the tree of shapes, $\overline{\Omega}$. Then the rest of both trees are separated into their maximal branches. In Fig. 5.2 we see how a branch of $\mathcal{L}(u)$ is linked to a node of $\mathcal{U}(u)$ according to the rules described above for the branches of $\mathcal{L}(u)$ (of type "$\mathrm{Sat}(B)$ is linked to $\mathrm{Sat}(N)$"). We see also how a maximal branch of

$\mathcal{U}(u)$ is linked to $\overline{\Omega}$ (which is the saturation of a node in both trees). In this case, it can be considered as a linking of type "Sat(B) is linked to Sat(B')" and also of type "Sat(B) is linked to Sat(N)". In Fig. 5.3 we see how a branch of $\mathcal{L}(u)$ is linked to another node of $\mathcal{L}(u)$ according to the rules described above for the branches of $\mathcal{L}(u)$ (of type "Sat(B) is linked to Sat(B')"). We see also how maximal branches of both trees $\mathcal{U}(u)$ and $\mathcal{L}(u)$ are linked to $\overline{\Omega}$. Observe that the trees of shapes obtained in both figures are equivalent.

Let S, T be two given shapes such that $S \subseteq T$. Let us prove that our construction of the tree of shapes by fusion of $\mathcal{U}(u)$ and $\mathcal{L}(u)$ has connected them. Observe that, since u is weakly oscillating, there are a finite number of leaves and a finite number of singular values. Hence, there is a finite number of singular shapes.

Lemma 5.7. *Let $S, T \in \mathcal{S}(u)$, $S \subseteq T$. Then there is a partition of $[S, T]$ into maximal monotone sections $\mathcal{M}_i$, $i = 1, \ldots, r$, such that $\mathcal{M}_i$ is at the left of $\mathcal{M}_{i+1}$ for each $= 1, \ldots, r-1$.*

Proof. Observe that given $Q \in [S, T]$ there is a maximal monotone section containing Q. Since there is a finite number of them, we have $[S, T] = \cup_{i=1}^{r} \mathcal{M}_i$ where $\mathcal{M}_i$ are the maximal monotone sections contained in $[S, T]$. We recall that each $\mathcal{M}_i$ is an interval of shapes which may be open or closed at its ends. Notice that $\mathcal{M}_i \cap \mathcal{M}_j = \emptyset$, for any $i \neq j$. Indeed, if $\mathcal{M}_i$ and $\mathcal{M}_j$ are of the same type, then by Lemma 2.28 they cannot intersect. Since the only shape of upper an lower type is $\overline{\Omega}$, if $\mathcal{M}_i$ and $\mathcal{M}_j$ are of different type and they intersect, then $\overline{\Omega}$ is the upper extrema of one of them, say $\mathcal{M}_i$, and the lower extrema of $\mathcal{M}_j$ (in which case $\mathcal{M}_j = \overline{\Omega}$). In that case, $\mathcal{M}_j$ would not appear since $\overline{\Omega}$ already appears in $\mathcal{M}_i$. Thus we may order them so that $\mathcal{M}_i$, $i = 1, \ldots, r$, constitute a partition of $[S, T]$ into intervals, $\mathcal{M}_i$ is at the left of $\mathcal{M}_{i+1}$ for each $= 1, \ldots, r-1$. □

Let us notice that a monotone section of upper (resp. lower) type of the tree of shapes is composed of saturations of nodes of a branch of $\mathcal{U}(u)$ (resp. of $\mathcal{L}(u)$). Indeed, if $[S, T]$ is a monotone section of $\mathcal{S}(u)$ and $S = \mathrm{Sat}(\mathrm{cc}([u \geq \lambda]))$, $T = \mathrm{Sat}(\mathrm{cc}([u \geq \mu]))$, $\lambda \geq \mu$, then the interval $[\mathrm{cc}([u \geq \lambda]), \mathrm{cc}([u \geq \mu])]_+$ is a branch of $\mathcal{U}(u)$ since it cannot contain any bifurcation of the upper tree. The presence of such a bifurcation would imply the presence of a bifurcation in the monotone section $[S, T]$, a contradiction. The same argument can be done for monotone sections of lower type.

Let us consider the interval $[S, T]$ in $\mathcal{S}(u)$. Let us consider two consecutive maximal monotone sections $\mathcal{M}$ and $\mathcal{M}'$ in $[S, T]$, the second one at the right of the first one. Let us consider four cases:

(*i*) $\mathcal{M}$ and $\mathcal{M}'$ are both of upper type. In that case, $\sup \mathcal{M} = \mathrm{Sat}(B)$ and $\inf \mathcal{M}' = \mathrm{Sat}(B')$ where $B \in \mathcal{CC}([u > \lambda])$ and $B' = \mathrm{cc}([u \geq \lambda]), B)$. Then there is a bifurcation at level λ and we are in the situation described in the fusion rules for $\mathcal{U}(u)$ where we linked Sat(B) to Sat(B').

(*ii*) $\mathcal{M}$ is of upper and $\mathcal{M}'$ of lower type. In that case, $\sup \mathcal{M} = \mathrm{Sat}(B)$ and $\inf \mathcal{M}' = \mathrm{Sat}(N)$ where $B \in \mathcal{CC}([u > \lambda])$ is a hole in $N \in \mathcal{CC}([u \leq \lambda]))$. Then there is a bifurcation at level λ and we are in the situation described in the fusion rules for $\mathcal{U}(u)$ where we linked $\mathrm{Sat}(B)$ to $\mathrm{Sat}(N)$.
(*iii*) $\mathcal{M}$ and $\mathcal{M}'$ are both of lower type. In that case, $\sup \mathcal{M} = \mathrm{Sat}(B)$ and $\inf \mathcal{M}' = \mathrm{Sat}(B')$ where $B \in \mathcal{CC}([u < \lambda])$ and $B' = \mathrm{cc}([u \leq \lambda]), B)$. Then there is a bifurcation at level λ and we are in the situation described in the fusion rules for $\mathcal{L}(u)$ where we linked $\mathrm{Sat}(B)$ to $\mathrm{Sat}(B')$.
(*iv*) $\mathcal{M}$ is of lower and $\mathcal{M}'$ of upper type. In that case, $\sup \mathcal{M} = \mathrm{Sat}(B)$ and $\inf \mathcal{M}' = \mathrm{Sat}(N)$ where $B \in \mathcal{CC}([u < \lambda])$ is a hole in $N \in \mathcal{CC}([u \geq \lambda]))$. Then there is a bifurcation at level λ and we are in the situation described in the fusion rules for $\mathcal{L}(u)$ where we linked $\mathrm{Sat}(B)$ to $\mathrm{Sat}(N)$.

We have shown that each transition between two consecutive maximal monotone sections of $[S, T]$ has been linked when fusing the upper and lower trees. The path from S to T in $\mathcal{S}(u)$ is reproduced in the fused tree.

Proposition 5.8. *The fusion of $\mathcal{U}(u)$ and $\mathcal{L}(u)$ according to the fusion rules produces $\mathcal{S}(u)$.*

When $N = 2$ the algorithm just described is less efficient than the algorithms described in Chaps. 6 and 7 but it is the most natural and can be easily implemented when $N \geq 3$. It was used in the first implementation of extraction of the tree of shapes for $N = 2$ in [79] and has been extended to 3D images by E. Meinhardt [20].

5.3 Fusion Algorithm in the Discrete Digital Case

We briefly sketch the fusion algorithm for a digital image in any dimension, though it is less efficient than our algorithm of choice for $N = 2$, the FLST, exposed in Chap. 6. The precise underlying function associated to a digital image will be made precise in that chapter. It is sufficient here to know that the function is piecewise constant and that the number of distinct shapes is finite. The procedure we describe follows closely the fusion rules for the continuous case.

Assume that the upper and lower trees have been computed. This can be done by starting at extrema and using region growing with priority queues (more details are given in Chap. 6). We describe here the fusion part.

We first remove all nodes that contain p_∞ in both trees and connect the resulting orphan nodes to the root of their respective tree.

For any connected component C of level set in $\mathcal{U}(u)$ or $\mathcal{L}(u)$, $\mathrm{Sat}(C)$ is not readily available but can be computed by following each component of the boundary of C and filling in. We actually do not need to do this computation, but we have to follow each component of the boundary of C to compute the number of pixels in $\mathrm{Sat}(C)$ by means of Green's formula. On top of that, if C is the upper limit of a maximal branch (either of $\mathcal{U}(u)$ or $\mathcal{L}(u)$), then we store one pixel in the exterior boundary of C.

Suppose to fix ideas that $C \in \mathcal{CC}([u \geq \lambda])$ is the upper limit of a maximal branch. We can consider its parent P in $\mathcal{U}(u)$ and the smallest node $D \in \mathcal{L}(u)$ containing the neighbor pixel of C. From D, we go up in the tree $\mathcal{L}(u)$ while the level remains strictly less than λ, call such maximal node still D. If the area of $\mathrm{Sat}(D)$ is less than area of $\mathrm{Sat}(P)$, we re-parent C from P to D, because $\mathrm{Sat}(C)$ is then a hole of D.

Chapter 6
Computation of the Tree of Shapes of a Digital Image

We present here a first algorithm, the Fast Level Sets Transform, that computes the tree of shapes from an input image. Although it is applicable in any dimension, its efficiency relies on specific properties of a dimension 2 domain. Although a direct derivation of the existence of the tree of shapes in the discrete case would be possible, a judicious interpretation of the digital image as a continuous domain image leads us directly in the framework of Chap. 2, hence proving the result. This artifice avoids the necessity of a theory in the discrete case. An open source reference implementation of the algorithm presented in this chapter is freely available in the MegaWave software suite (`http://megawave.cmla.ens-cachan.fr`, module `flst`).

6.1 Continuous Interpretation of a Digital Image

In order to apply the concept of saturation, holes and shapes to a digital image, we define a continuous image associated to the digital image. The resulting shapes of the continuous image can be mapped back to the digital image by considering the contained pixels.

6.1.1 From Digital to Continuous Image

Let us note u_d the digital image of dimensions W and H. It associates to pixel positions $(i,j) \in \Omega_d = \{0, 1 \cdots W-1\} \times \{0, 1 \cdots H-1\}$ a real gray level value $u_{i,j}$. Most of the time, the values are integers between 0 and 255 both included (coded in one byte), with 0 corresponding to black, 255 to white and values in between different shades of gray. In general, let us note u_{min} and u_{max} the minimal and maximal values of $u_{i,j}$.

First define a continuous image $\tilde{u}$ on the plane rectangle $[0,W] \times [0,H]$ by

V. Caselles and P. Monasse, *Geometric Description of Images as Topographic Maps*, Lecture Notes in Mathematics 1984,
DOI 10.1007/978-3-642-04611-7_6, © Springer-Verlag Berlin Heidelberg 2010

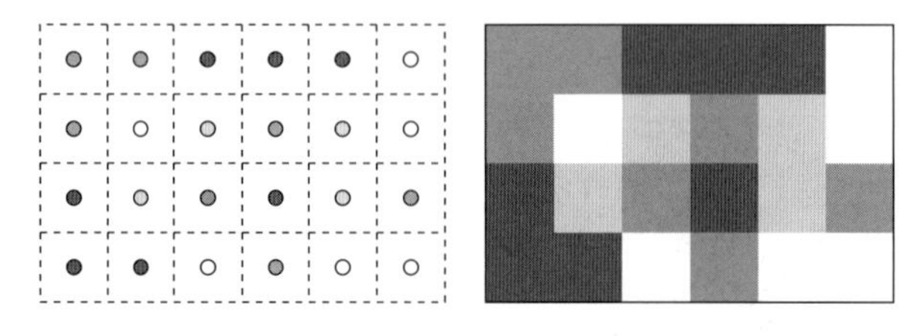

Fig. 6.1 The contrast-invariant interpolation we use to consider the digital image u_d (left) as a continuous domain image u_c (right). No new gray level is introduced and the values at edgels are such that u_c is upper semicontinuous.

$$\tilde{u}(x) = \begin{cases} u_d\left(x - \left(\frac{1}{2}, \frac{1}{2}\right)\right) & \text{if } x - \left(\frac{1}{2}, \frac{1}{2}\right) \in \Omega_d \\ u_{min} & \text{otherwise.} \end{cases}$$

Then consider the dilation u_c of $\tilde{u}$ by the closed unit square $B = [-1/2, 1/2]^2$ as structuring element:

$$u_c(x) = \sup_{y \in B} \tilde{u}(x + y).$$

A first remark is that u_c is an upper semicontinuous, contrast invariant interpretation of the digital image u_d, see Fig. 6.1. Indeed, $\tilde{u}$ is upper semicontinuous since its upper level sets are finite unions of isolated points (i, j), thus closed, and u_c is constructed from $\tilde{u}$ by dilation with a closed structuring element. The contrast invariant part derives from the facts that the extension of u_d to $\tilde{u}$ is contrast invariant and that the dilation is a morphological operator.

6.1.2 Digital Shapes

Digital shapes of u_d are deduced from the shapes of u_c by considering the pixels inside the continuous shapes. That is, we define a digital shape of u_d as the intersection of a shape of u_c with Ω_d.

The chosen extension of digital to continuous image answers for us the question of the choice of discrete connectivity: upper level sets must be partitioned by 8-connectivity and lower level sets by 4-connectivity. Digital shapes derived from upper shapes are 8-connected and those derived form lower shapes are 4-connected.

In this whole chapter, a pixel in the discrete setting means coordinates (i, j), while in the continuous setting it will be the part of the square $[i, i+1] \times [j, j+1]$ where u_c takes the value $u_d(i, j)$. In the latter case, a pixel can be open (local minimum), closed (local maximum), or neither.

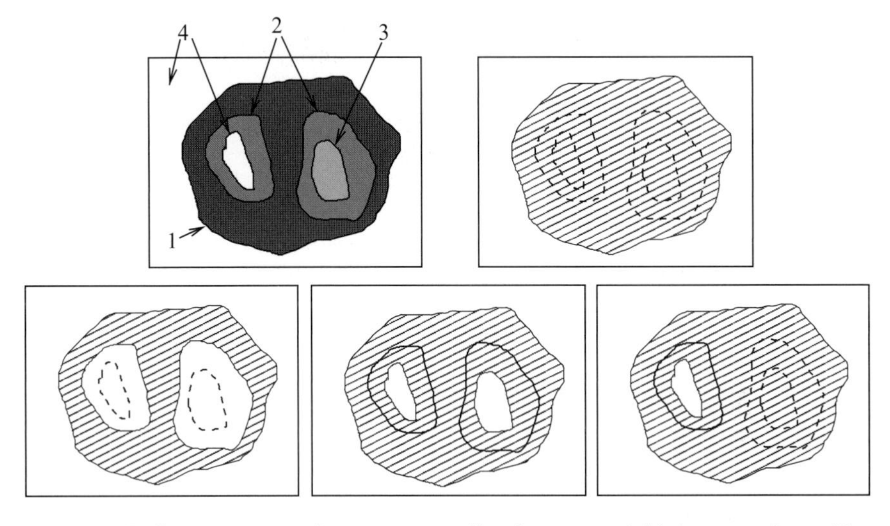

Fig. 6.2 Different connected components of level sets can yield the same shape. Top: an image u with gray levels indicated (left) and one of shape. Bottom, left to right: the connected components of lower level sets $\mathrm{cc}([u \leq 1])$, $\mathrm{cc}([u \leq 2])$ and $\mathrm{cc}([u \leq 3])$, all yielding the same shape.

For purpose of exact reconstruction, it is useful to associate a gray level to a digital shape, beside its type (upper or lower shape). First, only gray levels actually present in the image need be considered: an upper level set $[u \geq \lambda]$ can be written $[u \geq \lambda_d]$ with $\lambda_d = \min(u_d(\Omega_d) \cap [\lambda, +\infty))$, while a lower level set $[u < \mu]$ can be written $[u \leq \mu_d]$ with $\mu_d = \max(u_d(\Omega_d) \cap (-\infty, \mu))$. Second, we keep only the tightest level for a shape: it is possible for a connected component to grow only "in the holes" when changing the level, see Fig. 6.2. To an upper shape we associate the greatest level in Ω_d for which a connected component of the upper level set yields the shape by saturation. Similarly we associate the lowest level in Ω_d for which a connected component of $[u \leq \mu_d]$ yields a lower shape by saturation.

For a given pixel $P = (i, j)$ of value $u_d(P)$, we can consider the two shapes $S_P^+ = \mathrm{Sat}\,\mathrm{cc}([u_c \geq u_d(P)], P)$ and $S_P^- = \mathrm{Sat}\,\mathrm{cc}([u_c \leq u_d(P)], P)$, which are nested since they intersect at P. The smallest of both is the smallest of all shapes containing P. We call it the smallest shape associated to P.

Conversely, we can consider for a given shape S the family of pixels whose smallest associated shape is S. There is at least one of them. To see that, take the set of boundary pixels of a shape S, that is, the pixels in S for which at least one neighbor is not in S. "Neighbor pixels" are accounted for with the same connectivity as S, namely 8-neighbors if S is an upper shape and 4-neighbors if S is a lower shape. Then these pixels are connected (with the same connectivity as S) and it is easy to see that those among them at the level associated to S have S as smallest shape.

An important practical consequence of this is that the number of digital shapes is upper-bounded by the number of pixels.

6.2 Fast Level Set Transform

The FLST, for “Fast Level Set Transform”, is our algorithm to extract the shapes of a digital image and represent them in a tree structure.

6.2.1 Input and Output

As input, the FLST takes only a digital image. As output, it gives the tree of shapes. Let us detail the data structure used for it.

The values of the image are given in an array of size $W \times H$. It is convenient to enumerate it line after line, each line from left to right (as in text reading), we note the successive values $u_1, u_2 \cdots u_N$ with $N = W.H$.

A convenient data structure for a digital shape is:

Structure of a digital shape (node of the inclusion tree)
`Parent`, `Child`, `NextSibling`: Pointers to other nodes
`Type`: Lower or upper shape
`Gray`: The associated gray level
`Data`: Characteristics of the shape

The local tree structure is encoded in the shape by the fields `Parent`, `Child` and `NextSibling`. The root of the tree, i.e., the shape Ω_d has null parent, the leaves of the tree, i.e., regional extrema of the image without holes, have null child. The siblings (that is, different shapes having the same `Parent`) are encoded in a linked list, the link being given by `NextSibling`. In that manner, the procedure to enumerate the list of children of a shape is:

Input: The parent shape P
$C \leftarrow P$.`Child` /* C becomes successively each child of P */
while $C \neq \emptyset$ **do**
  Output C /* C is a child of P */
  $C \leftarrow C$.`NextSibling`

The field `Gray` of the shape is the gray level associated to the shape, as explained in Sect. 6.1.2. The field `Type` has no meaning for the root. The field `Data` can store additional data for the shape, such as area, perimeter, moments...

The inclusion tree of shapes has the following data structure:

Structure of the inclusion tree
W, H: Dimensions of the image
`Root`: The root shape of the tree
`SmallestShape` $= \{S_1 \cdots S_N\}$: Array of the smallest shapes

The array `SmallestShape` gives the smallest shape containing each pixel. For the pixel of index i, S_i points to its smallest including shape. The map $i \to S_i$ is called *component mapping* in the context of component tree [83].

6.2.2 Description of the Algorithm

Reconstruction The reconstruction algorithm is straightforward and of linear complexity, see Algorithm 1. Each pixel gets the gray value of the associated smallest shape. Notice that the `Type` of the shape is not used.

Input: The inclusion tree T
Output: Reconstructed image u
$u \leftarrow$Array of size $N = W.H$ `/* Creation of the array */`
for $i = 1$ **to** N **do**
  $u_i \leftarrow T.S_i$.`Gray`
return u

Algorithm 1: Reconstruction of the image from its inclusion tree

Basics of Extraction During the execution of the extraction algorithm, the image is modified. At the end, the image is uniform at the gray level of the root.

We use an image of markers, `Tag(`P`)` for a pixel P. Originally, `Tag(`P`)` is false for each pixel P; after a pixel is examined and its smallest shape found, it becomes true.

The neighbors of the current region are stored in a priority queue $\mathcal{N}$, the priority of a pixel depending on its gray-level. A priority queue is a data structure, supporting the following operations:

- push a new pixel P at priority g, $\mathcal{N}.enq(P, g)$;
- pop a pixel P at highest priority, $P \leftarrow \mathcal{N}.deq()$;
- return the top priority $\mathcal{N}.Priority$, that is either minimum $\mathcal{N}.\min$ or maximum $\mathcal{N}.\max$ priority.

The priority is the gray level of the pixel. Top priority may be either the minimal (for a lower level set) or the maximal gray level (for an upper level set). To be able to use the same structure for upper and lower level sets, a field $\mathcal{N}.Type$ (of value max or min) is added. At some point the expected type may be not yet known, in which case we add the value `Unknown`.

The operations above must be as fast as possible. Several implementations offer different performances. For example, a binary heap does both *enq* and *deq* in $O(\#\mathcal{N})$. For 8-bit priorities, a hash table does both in $O(1)$ at the cost of greater space requirement.

Main Loop First, initializations of variables take place: `Tag`(u_i) is set to false for all i and the tree T contains only the root.

The main loop of the algorithm is just a scan of the image, calling a procedure of extraction when a non-tagged local extremum of the image is found. The expected type of shape $\mathcal{N}.Type$ is set to min for a local minimum and max for a local maximum. The procedure `ExtractBranch` is explained below.

A pixel P is said to be a local extremum of u_d if and only if:

1. $\forall Q$ 4-neighbor of P, $u(P) \leq u(Q)$ (local minimum) or
2. $\forall Q$ 8-neighbor of P, $u(P) \geq u(Q)$ (local maximum);
3. the inequality is strict for one of these neighbors.

```
for i = 1 to N do
    if not Tag(u_i) and u_i is a local extremum then
        Set N.Type
        Call ExtractBranch(u_i)
```

Algorithm 2: The main loop of the FLST

Extraction of Branch We describe here the procedure `ExtractBranch`, which extracts the terminal branch at a given pixel P, if it exists. Compared to a branch (Definition 2.20), the terminal branch at P is the upper branch $\mathcal{B}_P$ provided S_P is a leaf.

Notice that the branch containing P can be empty.

This procedure extracts the terminal branch at P only if P belongs to a regional extremum without holes, that is a leaf of the tree. This is not a limitation, as Sect. 6.2.3 will show.

Definition 6.1. A subset R of Ω_d is said to be a regional minimum (resp. maximum) if it is a maximal subset for the relation "being connected and containing only local minima" (resp. maxima).

Remark 6.2. The notion of maximum and minimum is taken here in the large sense: if all neighbor pixels are at the same value, we say also it is an extremum.

Remark 6.3. Of course, this definition given for pixels, where the term "connected" means, as usual in this dissertation, 4-connected for minimum and 8-connected for maximum, remains identical for continuous domain images, with the topological definition of connectivity.

As expected, the regional extrema are disjoint.

Remark 6.4. A regional minimum (resp. maximum) is thus a connected component of isolevel for which all neighbors have a larger (resp. smaller) gray level.

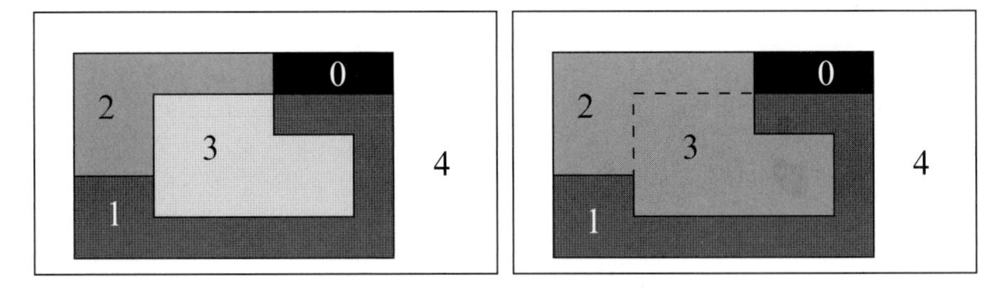

Fig. 6.3 Removal of a shape. Left: initial image. Right: image when the shape S at isolevel 3 is removed. S is a connected component of upper level set in the initial image, its pixels take the maximum gray level at its neighbors, 2.

`ExtractBranch` is a bottom-up, region growing algorithm. The idea is to remove a shape as soon as it is detected and stored in the tree. What we mean by removing a shape is putting all pixels of the shape at the lowest gray level of neighbors in case of a lower shape and at the largest gray level of neighbors for an upper shape (see Fig. 6.3).

As a shape containing P contains its connected component of isolevel, we need first a procedure to find this isolevel. The strategy is region growing, using the queue $\mathcal{N}$. This procedure, `ExtractParent`, is described in Algorithm 3.

Before adding a neighbor to $\mathcal{N}$, it must be checked that it is neither in R nor in $\mathcal{N}$ already (because it can be a neighbor of several pixels in R). A solution is to have an image of flags, `ImageNeighbors`, where each flag indicates if the pixel was added as a neighbor to the current region or not. To avoid reinitializing this image at each call of `ExtractParent`, we keep an index `Index` incremented at each call to `ExtractBranch`, and have for $\texttt{ImageNeighbors}_i$ a pair ($\texttt{Index}_i$,$\texttt{Flag}_i$). The flag can have one out of 2 values: `InN` or `InR`. The test to know whether a pixel was already considered as neighbor is "$\texttt{Index}_i = \texttt{Index}$" and the instruction to register the addition of a new neighbor is "$\texttt{Index}_i \leftarrow \texttt{Index}$ and $\texttt{Flag}_i \leftarrow \texttt{InN}$". The flag is switched to `InR` when the pixel is added to R. The interest of having a distinction between `InR` and `InN` is the use of the flag to compute the number of holes of the current set, as explained later.

The procedure `ExtractBranch` (see Algorithm 4) first extracts a regional extremum, stores it as a shape if it has no hole, extract the parent shape, etc. There are three break conditions:

1. The region is not a regional extremum;
2. The region has an internal hole;
3. The region either contains all boundary pixels or only some of them but contains more than half the pixels.

In the first two cases, the region is part of a bifurcation in the tree, that is there is another shape (not yet extracted) contained in it. In the third case,

```
Input: Region R, neighborhood N (both modified at output)
Output: Success status
R ← ∅                              /* The connected component of isolevel */
g ← N.Priority                                 /* The queue of neighbors */
while N.Priority = g do
|  Q ← N.deq()                                         /* Pixel to add */
|  for Q' 4-neighbor of Q do N.enq(Q', u(Q'))
|  if N.Priority = Unknown then                 /* Type discovery mode */
|  |  if g > N.Min then N.Type ← max; return false
|  |  if g < N.Max then N.Type ← min; return false
|  else Tag(Q) ← true; R ← R ∪ {Q}          /* Tag pixel Q and add to R */
|  if N.Priority = max then              /* We expect a regional maximum */
|  |  for Q' diagonal neighbor of Q do N.enq(Q', u(Q'))
|  if N.Min < g < N.Max then
|  |  return false                    /* Parent shape is a bifurcation */
return true
```

Algorithm 3: The procedure `ExtractParent` finds the parent of shape R, if it is not a bifurcation

```
Input: The pixel P
End ← false                          /* Flag of end of the procedure */
N.enq(P, u(P))                                    /* Initialization */
while End ≠ true do
|  g ← N.Priority
|  if not ExtractParent(R,N) or R has an internal hole then
|  |  End ← true
|  else if R ∩ frame ≠ ∅ and #R > N/2 then
|  |  Gray(Root) ← g
|  |  End ← true
|  else
|  |  Store R as new shape in tree
|  |  if N.Min = N.Max then
|  |  |  N.Type ← Unknown; ExtractParent(R,N)      /* Find N.Type */
|  |  |  End ← Restart
for Q ∈ R do u(Q) ← g;                   /* Leveling of R at level g */
if End = Restart then ExtractBranch(P)
```

Algorithm 4: The procedure `ExtractBranch`, extracting the (terminal) branch at pixel P

the root of the tree is reached, so that the level associated to the root is the current one (see Fig. 6.4).

Instead of removing the current regional extremum when it is found to be a shape (that is, set all its pixels to the nearest gray level of the neighbors), since we continue the region growing it is sufficient to make *as if* the shape was removed and consider that the pixels of the shape are at the current gray level. In the same manner, in the image `ImageNeighbors`, the flags of

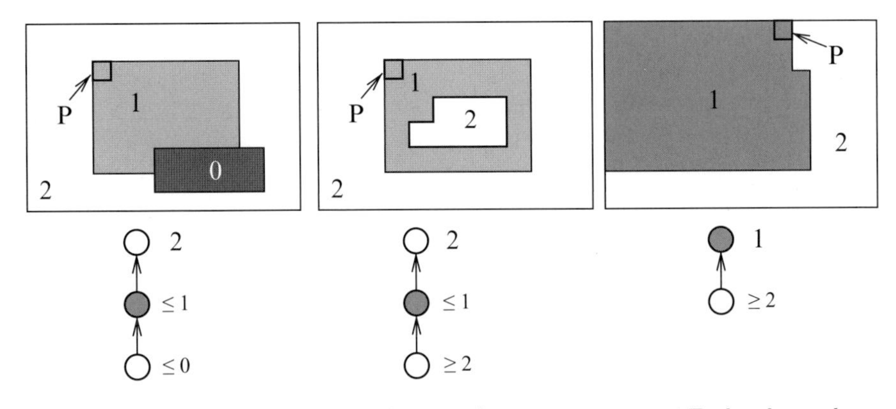

Fig. 6.4 The break conditions in the procedure `ExtractBranch`. Each column shows a configuration and its associated tree, with smallest shape containing the pixel P grayed. In each configuration, P is recognized as a local minimum. In the first case (left), P does not belong to a regional minimum. In the second case (middle), the smallest shape containing P is a regional minimum, but it has a hole. In the third case (right), the smallest shape containing P is a regional minimum, but meets the frame and has area larger than half the image, so it is the root.

the pixels of the found shape are already up to date, so nothing special is to be done in that image to continue the region growing[1].

After a shape is detected, it can happen that all its neighbor pixels are at the same gray level. In that case, the type of the parent is unknown, which has two consequences: we do not know which connectivity to use when resuming the region growing and we do not know whether $\mathcal{N}$ should be min- or max-oriented. The solution is to take 4-connectivity and look ahead by region growing in `ExtractParent` until a different gray level is found, which informs about the parent's type. Then `ExtractBranch` has to be recalled after removal of the branch, to keep correct track of the number of holes.

Choice of p_∞ A possible choice for the location of p_∞ (necessary choice to define the external hole of a set) is outside the image, then assume that u takes a constant value outside the image, $u(p_\infty)$, which can be a mean gray level, or some value outside of the range of the image. Although it has the desirable property that a level set not meeting image boundary is considered foreground, so the intuition of internal hole is preserved, it has the drawback that for a level set meeting the image boundary the background can be composed of several disconnected components of the complement within the image. A nice property would be to consider the external hole as always a single connected component of the complement. While a fixed position inside the image may be chosen, for example in one of the corners, we prefer to have

[1] That means that the `Index` used to avoid the reinitialization of `ImageNeighbors` must be incremented at each call of procedure `ExtractBranch`, not at the call of `ExtractParent`.

more invariance at least with respect to 90° rotations of image. With little effort, we can however adapt p_∞ to the image for this.

Given a connected set X containing at least one pixel, define the exterior of X, $Ext(X)$ as the connected component of $\overline{\Omega} \setminus X$ meeting the image boundary and comprising at least $\#\Omega_d/2$ pixels, or containing all image boundary pixels, if such a component exists, in which case it is obviously unique. The rationale for the first condition of the definition is that we consider as background (exterior) the component occupying the major part of the image. For convenience, we take $Ext(X) = \overline{\Omega}$ if no such component exists (instead of the logical $\emptyset$). Given two connected components of upper (resp. lower) level sets X and Y (possibly at different levels), they are either nested or disjoint. If they are nested, it is easy to see that their exteriors are nested. We show that it is possible to choose a p_∞ in all exteriors of connected components of level sets.

We define the set

$$\Omega_\infty = \bigcap_{\lambda \in \mathbb{R}, X \in \mathcal{CC}([u \geq \lambda])} Ext(X) \cap \bigcap_{\lambda \in \mathbb{R}, X \in \mathcal{CC}([u \leq \lambda])} Ext(X).$$

We show that in general $\Omega_\infty \neq \emptyset$ and $\#u(\Omega_\infty) = 1$, that is all elements of Ω_∞ have same value. Then any choice of $p_\infty \in \Omega_\infty$ yields the same tree.

By ignoring all components such that $Ext(X) = \overline{\Omega}$ and taking maximal components in resulting upper and lower trees, we can write $\Omega_\infty = \bigcap_{i=1,\dots,m} Ext(X^i) \cap \bigcap_{i=1,\dots,n} Ext(X_i)$, X^i denoting connected components of upper type and X_i of lower type, with $X^i \cap X^j = \emptyset$ and $X_i \cap X_j = \emptyset$ if $i \neq j$.

According to Lemma 2.2, $K = \bigcap_{i=1,\dots,n} Ext(X_i)$ is a continuum. But since $Ext(X_i)$ is a hole of X_i, it can also be written as $\mathrm{Sat}(Y^i)$, Y^i being a component of upper type, for some choice of the point at infinity (that can depend on i). Therefore any X^i is either inside all $\mathrm{Sat}(Y^j)$, or outside one of them. The latter case does not change the intersection, so we can assume all X^i to be inside. We can then write

$$\Omega_\infty = K \setminus \bigcup_i \mathrm{Sat}(X^i),$$

each saturation written with respect to its own point at infinity. In any case, $\mathrm{Sat}(X^i)$ are disjoint continua in K, so that if there are two or more of them, their union is not a continuum and $\Omega_\infty \neq \emptyset$. If there is only one, X^1, the only possibility that $K \subseteq \mathrm{Sat}(X^1)$ is that $\mathrm{Sat}(X^1)$ meets the boundary of the image and contains exactly $\#\Omega_d/2$ pixels. In other words, the only case where $\Omega_\infty = \emptyset$ is when a level line meeting the image boundary separates the image in half.

Let p_1 and p_2 be two points of Ω_∞. Consider the set $X = \mathrm{cc}([u \geq u(p_1)], p_1)$. We have $p_1 \in X \cap Ext(X)$, which implies $Ext(X) = \overline{\Omega}$. Suppose

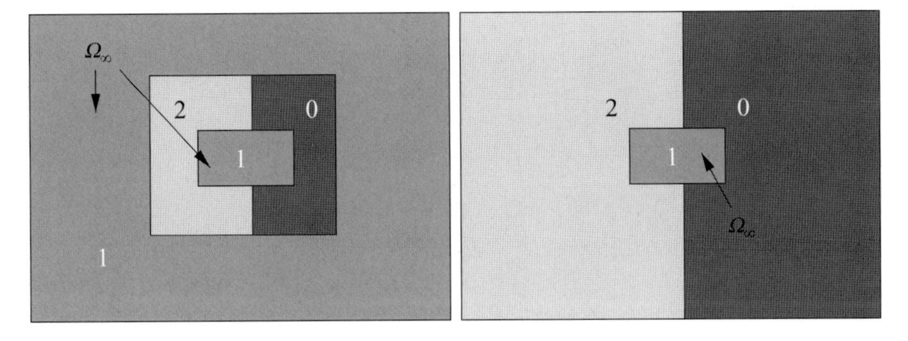

Fig. 6.5 Location of p_∞. Left image shows Ω_∞ may not be connected and right image that Ω_∞ may not even meet the frame boundary.

p_2 is in a hole H of X. Then we can write $H = \mathrm{Sat}(Y, p_1)$ with $Y \in \mathcal{CC}([u < u(p_1)])$. It is easy to check that $Ext(H) \neq \overline{\Omega}$. But since $p_2 \in \Omega_\infty \subseteq Ext(H)$, we get a contradiction with $p_2 \in H$. Therefore $p_2 \in X$, proving $u(p_2) \geq u(p_1)$. By symmetry of the roles of p_1 and p_2, we have the other inequality and finally $u(p_2) = u(p_1)$.

The value taken by u on Ω_∞ is the value of the root of the tree. Notice that Ω_∞ is not shown to be connected, and indeed it can be disconnected, see Fig. 6.5.

In the rare case where $\Omega_\infty = \emptyset$, we can still have an implicit choice of p_∞ by duplicating it: Since there is a level line separating the image in two halves, we can take a p_∞ in one half to extract the tree of the other half and vice versa. We have thus two trees, which can always be children of a root having no private pixel. Private pixels of a shape are its pixels that are not in any child.

Counting Holes We compute the number of internal holes in a regional extremum during the region-growing. Since we use 8-connectivity (resp. 4-connectivity) for this set and 4-connectivity (resp. 8-connectivity) for its complement, the Euler characteristic can be computed from *local* configurations of the set. This is proved by Rosenfeld and Kak in [88]. Notice that this property is lost when the same notion of connectivity is used for the set and its complement (see Kong and Rosenfeld [49]).

We recall that the Euler characteristic of a set S is $n - m + 1$ where n is the number of connected components of S and m the number of connected components of its complement. Since we deal with connected sets, $n = 1$, so that the Euler characteristic is $1 - h$ where $h = m - 1$ is the number of internal holes of S. The local configurations in question are blocks of at most 2×2 pixels (see Fig. 6.6). We count the variation of the Euler characteristic of a set when we add one pixel. This means that we have to examine the 3×3 neighborhood centered at the new pixel to compute this variation, since the

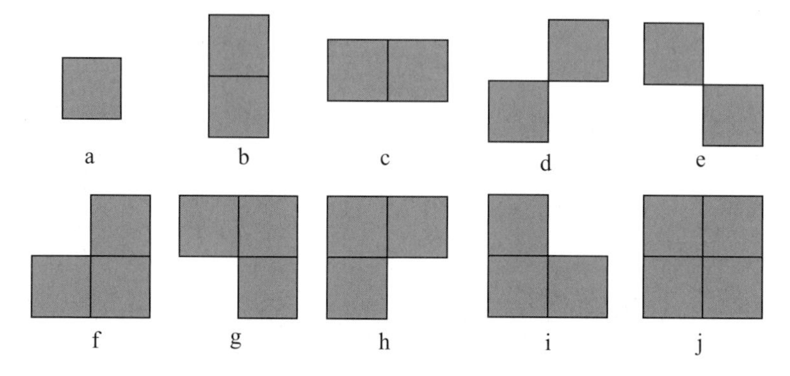

Fig. 6.6 The Euler characteristic of a set S of pixels in 4-connectivity (resp. 8-) and complement in 8-connectivity (resp. 4-) can be computed by counting the number of configurations on blocks of size at most 2×2. The formula is $a - b - c + j$ (resp. $a - b - c - d - e + f + g + h + i - j$), the letter indicating the number of possible positions for corresponding dominos in S. Notice $a = \#S$, the cardinality of S.

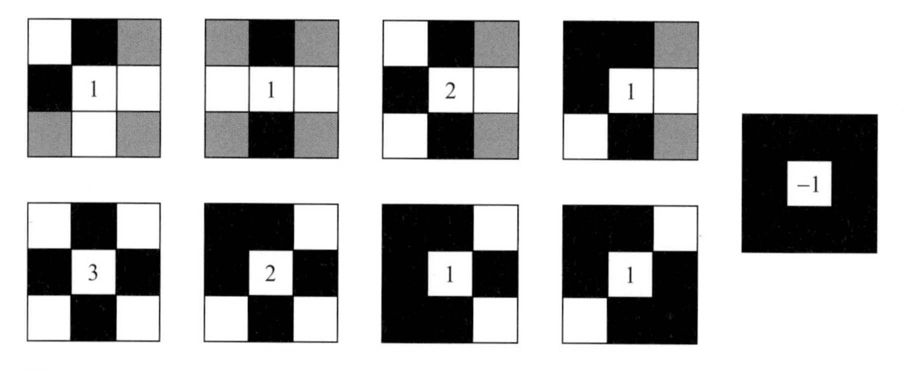

Fig. 6.7 The variation of the number of holes of a shape when we add a 4-neighbor P to a 4-connected set S, according to the local configurations, that is, the 3×3 neighborhoods centered at P. Pixels of S are represented in black, pixels not in S are left in white and pixels being in S or not, indifferently, in gray. To these configurations, we must add all their quarter and half turn rotates and the vertically or horizontally symmetric configurations. The configurations which cannot be obtained by such operations either do not modify the number of holes or are impossible (P would not be a 4-neighbor of S).

only 2×2 changed blocks are included in this 3×3 neighborhood (see Figs. 6.7 and 6.8).

When the added pixel is at the boundary of the image, the 3×3 neighborhood strides the image boundary, so should the pixels outside be considered inside or outside the region? The answer is: outside the first time the region meets the image boundary, inside after that. Indeed, consider the set $\tilde{R}$, union of a region R and all pixels outside the image. It is clear that R and $\tilde{R}$ have the same number of holes. But R has always one component while $\tilde{R}$ has two components if R does not meet the image boundary and one

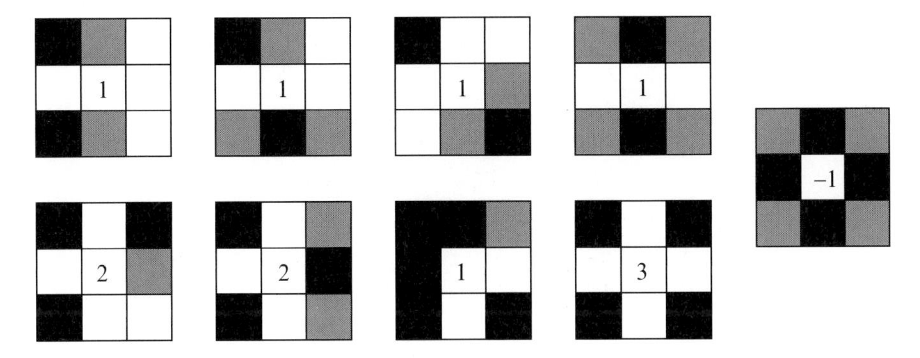

Fig. 6.8 The variation of the number of holes of a shape when we add an 8-neighbor P to an 8-connected set S, according to the local configurations around P. The rotated and symmetric versions of these configurations add or subtract the same number of holes. See Fig. 6.7 for explanations about the configurations.

component otherwise. So R and $\tilde{R}$ have same Euler characteristic only when R meets the image boundary. Now, if R has only one boundary pixel, the Euler characteristic is the same whether pixels outside are considered in or out[2]. That is why we switch from considering external pixels to be outside (when adding the first image boundary pixel) to inside thereafter.

A problem arises when we do not know the connectivity to choose, because all neighbors of the current set are at the same gray level. As explained above, we use in this case 4-connectivity for the set (and therefore 8-connectivity for the complement). But if we find a neighbor of strictly lower gray level, we switch to 8-connectivity. This can induce errors in the computation of the Euler characteristic, see Fig. 6.9. That explains why we restart the procedure `ExtractBranch` after having discovered the correct connectivity to adopt thanks to `ExtractParent`.

Storing Shapes The last point to explain is how we store the shapes when they are detected. That is, we need to find its place in the tree. This is done with the help of the array `SmallestShape`, and we explain also how to fill this tree. For the sake of performance of the algorithm, we also use an array `LargestShape`, which becomes useless at the end of the FLST. This array, whose elements we note $L_1, \cdots, L_N$ are pointers to the largest extracted shape (except Ω_d) containing each pixel in the tree. Of course, this array can be deduced from `SmallestShape` and the tree: It is enough to start from S_i and go up the tree until we reach a node whose parent is the root; this node is L_i. Nevertheless, having this array in memory saves computation time.

Initially, we set each S_i to $\emptyset$ and each L_i to `Root`. When a new shape S is detected, we have its set of pixels R. For each pixel u_i of R whose associated S_i is $\emptyset$, we set S_i to S. For each u_i of R whose associated L_i is not `Root`,

[2] Except when the image is 1 pixel wide or high, a case easily dealt with.

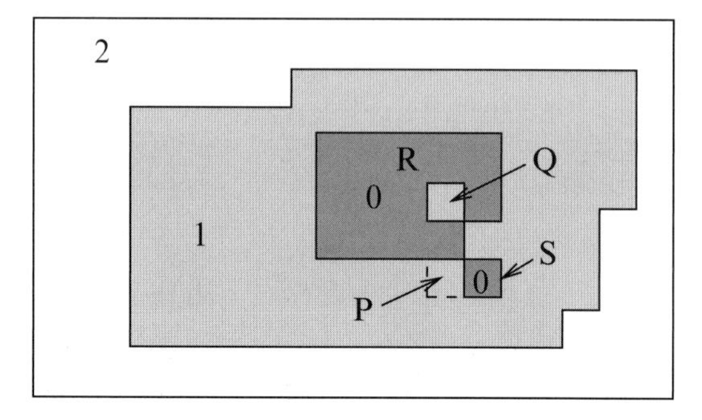

Fig. 6.9 Possible error in Euler characteristic when all neighbors of the current set are at the same gray level. The region R is a connected component of lower level set, at level 0, so considered in 4-connectivity, without hole. When the shape is removed, all neighbors are at same gray level, 1. If the neighbor P is first appended to the current region, it has S as neighbor, at a lower gray level, 0. So we switch to 8-connectivity. But this modifies the Euler characteristic of the current region, which becomes 2, because the pixel Q is now a hole in the current region. When Q is appended to the region, this decrements the Euler characteristic, which must be corrected to be 1 after Q is appended.

we put the shape pointed to by L_i as child of S, the subtree rooted at the node L_i remaining unchanged. Finally, for each pixel $u_i \in R$, we set the associated L_i to S.

6.2.3 Justification of the Algorithm

This section explains why the presented algorithm succeeds in extracting the tree.

An Internal Hole of a Level Set Is a Shape This is a direct consequence of the last assertion of Lemma 2.12 applied to $X = [u \geq \lambda]$ or $X = [u \leq \lambda]$, which indeed have a finite number of components since they are composed of pixels. A leaf of the tree is thus a regional extremum without internal hole. That explains why the procedure `ExtractBranch` aborts when an internal hole is detected.

The Level of the Parent of a Shape Is the Closest Level of Its Neighbors If a shape S is at level λ, then the parent T is at the level μ closest to λ among the neighbors of S. Let P be a neighbor pixel at this level μ.

Let us write $S = \text{Sat}(X)$ and $T = \text{Sat}(Y)$ with X and Y connected components of level sets, and N the union of neighbor pixels of S. Since $S \cup N$ is connected, it cannot be in an internal hole of Y, which would be a shape between S and T. Therefore $Y \cap N \neq \emptyset$, meaning that Y contains a neighbor of S.

If X and Y are of the same type, then $X \subseteq Y$ and $P \in Y$, thus the level of Y is μ.

If X and Y are of different types, S is an internal hole of Y. If some neighbor Q were not in Y, $S \cup Q$ would be in a shape strictly inside Y, a contradiction. Therefore $N \subseteq Y$. Since N is connected and S is a hole of N, the level of Y is μ.

Correctness of the Algorithm The above results show that:

- Each branch can be removed by removing successively terminal shapes;
- There is no need to make more than one scan of the image, since during a scan, the associated branches of all scanned pixels have been extracted;
- After a complete scan of the image, there remains no branch, so all shapes of the image have been extracted.

Example Figure 6.10 presents an example of execution of the algorithm. Each step presents the image, the (unknown) inclusion tree, and the part of the inclusion tree currently computed. The steps are the following:

(a) The original image. The computed tree is initialized with only the root, whose gray level is unknown.

(b) The first local minimum encountered during the scan is pixel P. The procedure `ExtractBranch` is called, it finds the isolevel 2, but it is not a regional extremum, so it exits. Pixel Q is the next local extremum, this extracts isolevel 5, which is a regional extremum, but with holes, so continue the scan. The same situation happens for R. S is the next local extremum. The procedure `ExtractBranch` first finds the regional extremum at level 5 containing S, it has no hole, so it adds it in the computed tree and removes it from the image.

(c) `ExtractBranch`, called from pixel S continues the region growing and extracts the regional extremum at level 3, without hole, so adds it in the computed tree. When it grows the region further, it finds the regional extremum at level 2, but with one hole, so it exits.

(d) The next local extremum met during the scan is T. `ExtractBranch` finds the regional extremum at level 1, adds it in the tree, then finds the regional maximum at level 5, which has holes, so exits.

(e) U is the next local extremum, but its component of isolevel is not a regional extremum. Then V is met, but its associated regional extremum has a hole. On the contrary, at W a regional minimum without hole at level 0 is extracted.

(f) The region growing initiated at W continues and `ExtractBranch` finds the regional maximum at level 4.

(g) `ExtractBranch` continues the region growing and finds a regional extremum meeting the frame but with area less than half that of the image. So it is a shape. The following extracted isolevel has a hole, so `ExtractBranch` exits.

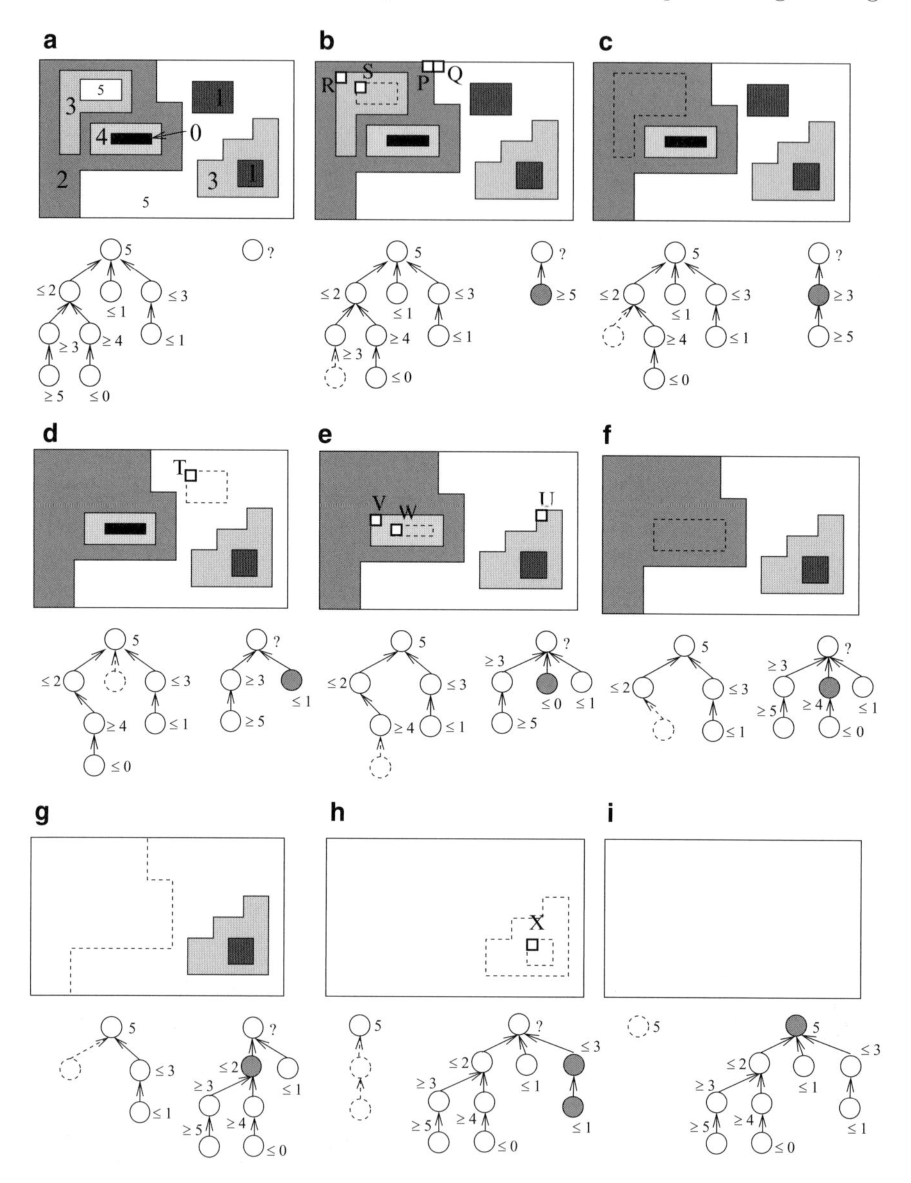

Fig. 6.10 An example of execution of the FLST at different stages, **a** to **i**. At each stage, the tree of the current image and the partially computed tree are shown. See text for details.

(h) The next non tagged local extremum is X, `ExtractBranch` finds successively the inferior shapes at level 1 and 3.

(i) When the region growing continues, `ExtractBranch` finds the whole image. This gives the gray level of the root, 5.

6.2.4 Complexity

Extraction of the tree relies on comparison of values of pixels. Thus the measure of complexity is the number of comparisons of values of pixels.

A pixel having 4 or 8 neighbors, we have to count how many times it is added to a region R in Algorithm 3. It could be several times as the procedure `ExtractBranch` can fail due to one break condition encountered. The break condition is essentially that a bifurcation is met in the region growing procedure. The number of passes through a given pixel is thus linked to the number of bifurcations among its containing shapes and the number of children of the smallest containing shape.

Let us consider the two toy examples in Fig. 6.11. For a shape S_i we note:

- p_i its number of private pixels.
- c_i its number of children.
- s_i its number of siblings, uncles, great-uncles, etc.

To be precise, s_i is the number of shapes that are not ancestors of S_i but whose parent is an ancestor of S_i.

When algorithm is applied to the left image, all private pixels of the root (at level 0) are first found, until finally an internal hole is found in the resulting region. The same happens for pixels at level 1, and so on until we reach pixels at level k. At this point, the upper shape k is extracted, then $k-1$, until 0. We see that each pixel is included twice in a region (first time resulting in a break condition, the second time while extracting the branch), except pixels at level k which are included once (no break condition). The complexity is thus $O(2p_0 + 2p_1 + \cdots + 2p_{k-1} + p_k)$, which amounts to $O(N)$ since $\sum_i p_i = N$.

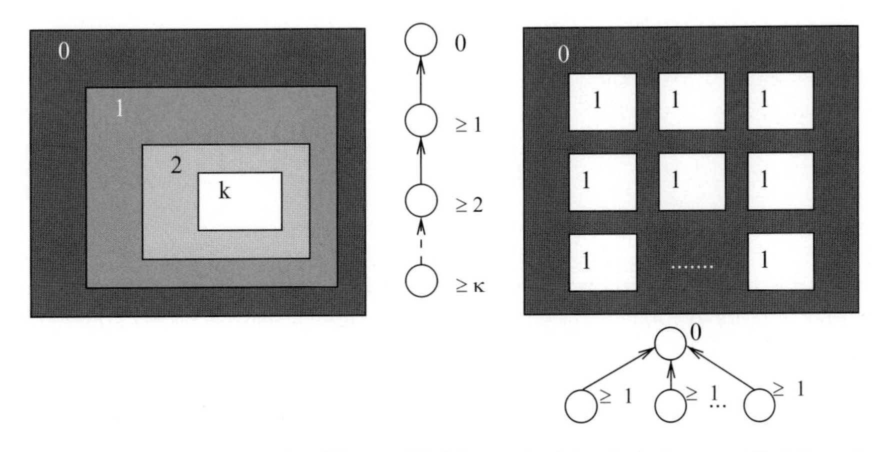

Fig. 6.11 Toy examples for different FLST complexities. Left: image with k linearly nested shapes $O(N)$. Right: all k shapes are children of root $O(N^2)$.

When applied to the right image, the image first scans the p_0 private pixels of the root, and ExtractBranch stops since there are k internal holes. Then the first shape at level 1 is found (p_1 pixels), adds the p_0 private pixels of root and finds $k-1$ internal holes in the resulting region. The second shape at level 1 is found (p_2 pixels) and then a region of $p_0 + p_1$ pixels at level 0 is detected (since the first shape at level 1 was leveled after extraction) as having $k-2$ holes. Finally, when reaching the kth shape at level 1, all other shapes have been leveled at 0, and the algorithm finishes. The number of comparisons was thus proportional to

$$p_0 + (p_0 + p_1) + (p_0 + p_1 + p_2) + \cdots + (p_0 + p_1 + \cdots + p_k).$$

We can also write it

$$p_0(k+1) + p_1 k + p_2(k-1) + p_3(k-2) + \ldots$$

If we assume $p_i = 1$ for $i = 1, \cdots, k$, we get

$$(N-k)(k+1) + \frac{k(k+1)}{2}.$$

If furthermore we assume that the shapes are just one pixel apart, we have $k = N/4$, so that the overall complexity is $O(N^2)$.

Comparing these two examples, we see that the same number of shapes yields radically different complexities depending on their organization, between $O(N)$ and $O(N^2)$. Monotone sections with many shapes are not a problem while bifurcations have a dramatic impact because they break the region growing in ExtractBranch as many times as there are children.

We see that a private pixel of shape S_i can be included in a region one time when a local extremum is detected, c_i times when it is reached by region growing through the children, and s_i times because a sibling, uncle or great-uncle breaks the region growing at the common ancestor. Therefore the complexity of the FLST is bounded by:

$$O\left(N + \sum_i p_i(c_i + s_i)\right).$$

If there is no bifurcation in the image, the complexity is simply $O(N)$, linear in the number of pixels. However the worst case, a checkerboard similar to Fig. 6.11 (right) can reach a complexity $O(N^2)$. For a moderately large image with $N = 10^6$, such a pathological case makes the algorithm excruciatingly long. Notice that a similar effect can happen in practice with a noisy photograph of the sky, a large uniform surface (sky) with many included small shapes (noise).

Image	Gaussian noise	texture	checkerboard	Lenna

Size	512 × 512	2400 × 1600	64 × 64	256 × 256
Comp./pixel	10.0	11.4	1512	11.3
Multi-pass	9.8	10.7	5.6	7.0

Fig. 6.12 Comparisons per pixel for different images with FLST and multi-pass variant.

This penalty can be amortized with a multi-pass algorithm: if we put a maximum area a (typically $a = 20$) as an additional break condition in `ExtractBranch`, we do not "pay" p_i pixels for each bifurcation under S_i but only $\min(p_i, a)$. In the first pass, only small shapes are extracted and erased from the image. Increasing the threshold a and doing a second pass we can extract bigger shapes. Given the fast decrease of the number of shapes relative to area in natural images, the threshold a can be modified according to a geometric progression between passes, for example 20, 200, 2000, N for $N = 10^6$, since the number of possible terminal branches of at least 2000 pixels cannot be above 500.

Figure 6.12 shows the number of comparisons of levels per pixel for different types of images. Notice that the multi-pass variant never degrades the number of comparisons, and actually provides an improvement more than two orders of magnitude for the worst-case checkerboard image. With this variant, the number of comparisons remains moderate, of the same order of operations as comparing each pixel to its 8 neighbors.

As a reference, consider the extraction of the connected components of level sets. To simplify, omit the extraction of holes. The basic algorithm applies successively each possible threshold to the image and applies each time a linear time algorithm to extract the connected components of black and white pixels in the binary image. Each pixel must be compared to its north and west neighbors (and north-east neighbor in case of 8-connectivity), which amounts to 2 or 3 comparisons per pixel. The complexity is thus $2g.N$ where g is the number of gray levels in the image and N the number of pixels. It is apparently linear $O(N)$, but with a constant at least $2g$ before, which can be typically 200, and at most (if all gray levels are used) $2 \times 256 = 512$. This is much higher than our experimental results with the FLST. Moreover, if gray levels are represented with floating point values, at worst each pixel has a unique gray level and $g = N$, therefore the basic algorithm becomes of complexity $O(N^2)$. This complexity makes it too slow for practical applications, whereas the FLST remains as efficient.

6.2.5 Comparison to Component Tree Extraction

Several algorithms for the computation of component tree were proposed in the literature. The most cited algorithm is by Salembier, Oliveras and Garrido [97]. However, its complexity is quadratic in the number of pixels if all pixel values are different: It is rather designed for 8-bit images. Improvements based on a kind of watershed algorithm were proposed by Najman and Couprie [83]. This algorithm is almost linear, provided the pixels can be sorted in linear time. This is easy to achieve for 8-bit images and almost achievable for fixed-size floating point values, where the sort can be made in $O(n \log \log n)$. Then the algorithm relies on Union-Find procedure of Tarjan [111], which is an efficient data structure to deal with a collection of disjoint sets. The method of [83] has to manipulate two such structures, one handling components at a given level and another for tree construction. Whereas the algorithm has quasi-linear complexity, it is reported in [83] to be slower than [97] on 8-bit images. A specialized algorithm for 8-bit images by Nistér and Stewénius [85] has a linear complexity and is particularly efficient. It relies on a flooding algorithm that does not require sorting the pixels, and computes one component at a time, avoiding the overhead of a Union-Find procedure. Upon analysis, however, the algorithm of [85] is an unacknowledged non-recursive implementation of the original algorithm of Salembier *et al.* [97]. Let us finally mention the algorithm by Berger *et al.* [13] which computes the component tree in an efficient way and is adapted both to images with fine and coarse quantization. The paper [13] also contains a comparison of several of the different trees we just mentioned.

Clearly, the FLST, if simplified for component tree computation, cannot compete with these methods. However, it is still fairly efficient (see Sect. 6.2.4), does not require 8-bit images, and builds the tree of shapes directly. For that goal, it is more efficient than computing the two component trees and merging them, as exposed in Chap. 5.

6.3 Taking Advantage of the Tree Structure

The fact that we have an inclusion tree structure permits to compute with a small amount of memory or computation time some characteristics of the shapes. We show two examples of such possibilities. The first one illustrates the economy of memory, the second one of computation time.

6.3.1 Storage of Pixels of the Digital Shapes

We show here how we can store the pixels of all shapes with a high economy of memory. The FLST gives for each shape, once it is detected, its list of pixels. But storing for each one this list would be memory greedy, because shapes can be nested, so that each pixel appears in different (maybe numerous) lists. This can result in the need of a high amount of memory, consequently larger than the size of the image, N. We propose instead a method where each pixel is listed only once.

For this, the shapes need to be enumerated in preorder, meaning that each shape is enumerated before its children shapes and siblings are enumerated adjacently (see Sedgewick, [101]). This is not the order in which the FLST extracts the shapes. Whatever the initial order of the shapes, reordering them is an $O(N)$ process. This can be done by a recursive procedure, but the non recursive version is also simple to implement, see Aho *et al.*, [1]. We suppose that the shapes are enumerated in such an order, name them $\{T_1, \dots, T_k\}$, where k is the number of shapes, such that

1. $\forall i, j, \quad T_i \subseteq T_j \Rightarrow i \geq j$;
2. $\forall i, j, l, \quad T_i \subseteq T_l$ and $T_j \subseteq T_l \;\Rightarrow\; \forall k$ between i and j, $T_k \subseteq T_l$.

Once this is done, we store for each shape its number of private pixels. We call the private pixels of a shape S the pixels u_i whose smallest containing shape S_i is S. It was proved that each shape contains at least one private pixel. This number is the area of S minus the sum of the areas of its children, so computed in $O(k)$, which is at most $O(N)$. We note these numbers p_i. The area of the shapes is given by the FLST during extraction and can be stored, or it can be recovered after extraction (see the following section).

The idea is to enumerate the pixels in an array A with an order induced by the preorder of the shapes, so that pixels belonging to the same shape are in an interval in A. Then each shape T_i needs only to store the index I_i of the beginning of the interval in A (and the length of the interval, that is, the area of the shape). This is explained in Algorithm 5. Each such interval contains the private pixels of the shape, then the pixels of its children. For example, the first p_1 pixels are the private pixels of the root, the following p_2 pixels the private pixels of the first child of the root, etc. The complexity of this algorithm is obviously $O(k)+O(N) = O(N)$ and the amount of memory necessary $O(N)$.

6.3.2 Computation of Additive Shape Characteristics

We can also take advantage of the tree structure to compute some additive characteristics associated to shapes. We call additive characteristics of a set a real number c such that for any two sets S_1 and S_2

Input: T_i and p_i, for $i = 1 \dots k$: Shapes in preorder and their number of private pixels
Input: S_j, for $j = 1 \dots N$: smallest containing shape of pixels
Output: $(A_j)_{j=1\dots N}$ array of sorted pixels, $(I_i)_{i=1\dots k}$ index in A of first pixel of shape i

```
l ← 1                                    /* Initialization of the current index */
for i = 1 to k do                        /* Loop of computation of indices I_i */
    I_i ← l
    l ← l + p_i
Copy array (I_i) in array (I'_i)   /* I'_i is the first free spot for a private
pixel of T_i */
for j = 1 to N do                                /* Enumeration of the pixels */
    i ← index such that S_j = T_i       /* Index of shape pointed to by S_j */
    A_{I'_i} ← j                                 /* Add pixel j in array A */
    I'_i ← I'_i + 1                                       /* Increment I'_i */
```

Algorithm 5: Efficient storage of the pixels associated to each shape

Input: $\{T_1, \dots, T_k\}$, the shapes in preorder
Input: $\{S_1, \dots, S_N\}$, smallest containing shape of each pixel
Output: $\{c_1, \dots, c_k\}$, additive characteristics

```
c_i ← 0 for i = 1...k                                      /* Initialization */
for j = 1 to N do                          /* Loop concerning private pixels */
    i ← index of S_j, such that S_j = T_i  /* Index of shape pointed to by S_j
    */
    c_i ← c_i + c(j)   /* c(j) is the characteristic of the set constituted
    of pixel number j */
for i = k down to 2 do      /* Order is important, so characteristics of
children are computed before their parent */
    i' ← index of parent of T_i
    c_{i'} ← c_{i'} + c_i     /* Add characteristic of child to characteristic of
    the parent */
```

Algorithm 6: Computation of an additive characteristic c of the shapes

$$S_1 \cap S_2 = \emptyset \quad \Rightarrow \quad c(S_1 \cup S_2) = c(S_1) + c(S_2).$$

Examples of this are numerous. In particular the area, more generally the moments, and any integral quantity. On the contrary, some other interesting characteristics are not, such as the perimeter.

We can compute additive characteristics of all shapes with linear complexity $O(N)$. This is not straightforward, since shapes can be nested.

Again, the idea is using the inclusion tree structure. Assume as in the previous section that the shapes are enumerated in preorder in an array $\{T_1, \dots, T_k\}$. Then the algorithm is in two steps:

1. Compute the characteristics of the set of private pixels of each shape.
2. Add the characteristic of each shape to its parent.

Indeed, since a shape is composed of its private pixels union the family of the private pixels of all its contained shapes, we are sure we describe all the shape. The method is described in Algorithm 6. The complexity is evidently of order $O(N)$ additions.

6.4 Alternative Algorithms

6.4.1 Changing Connectivity

The algorithm was presented for a discrete image considered as a sampling of a continuous domain image. This allows to use topological results in the plane and translate them in the discrete framework. In particular, the question of the connectivity is central in the algorithm. It appeared that using 4-connectivity for one type of level set and 8-connectivity for the other type corresponds implicitly to choosing the associated continuous domain image upper or lower semicontinuous. However, it is legitimate to wonder if the use of only *one* notion of connectivity leads also to an inclusion tree of shapes.

If we work only with k-connectivity ($k = 4$ or 8), are two intersecting shapes nested? The shapes in this case are the k-connected components of level sets, union their holes, where the holes are k-connected components of the complement.

When $k = 4$, which is what M-connectedness (as defined in [6, 10]) assumes, the answer is negative. Figure 6.13 shows a configuration where two shapes are intersecting, whereas none contains the other one. There is an easy explanation of this fact: the two pointels necessary to the connection of the two 4-connected components of the isolevel $[u = 0]$ have a value less than 1, so that the connected component of upper level set $[u \geq 1]$ does not connect to the background at level 2. Similarly, they must have a value more than 1, to prevent the connection with the other connected component of $[u = 0]$. The requirements on the values of these pointels are contradictory. The solution is to remove these pointels from the definition set of the image. That is to say, considering 4-connectivity for sets as well as for their complement amounts to imagine the associated continuous domain image defined on $\bar{\Omega}$ minus the junction pointels. The problem with this definition set is that it is not unicoherent, hindering the representation of the image as an inclusion tree of shapes.

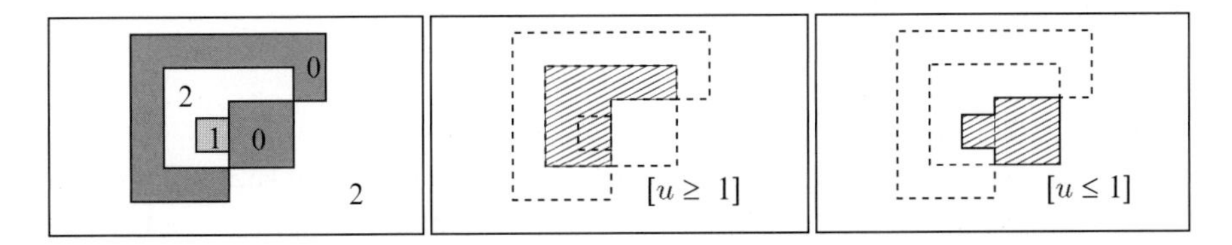

Fig. 6.13 A configuration where the choice of 4-connectivity for sets and also for their complement does not yield an inclusion tree of shapes. Left: the image u. Center: the connected component of upper level set $[u \geq 1]$ has no hole, it is a shape. Right: the connected component of lower level set $[u \leq 1]$ has no hole either, it is also a shape. Nevertheless, these two shapes intersect without being nested.

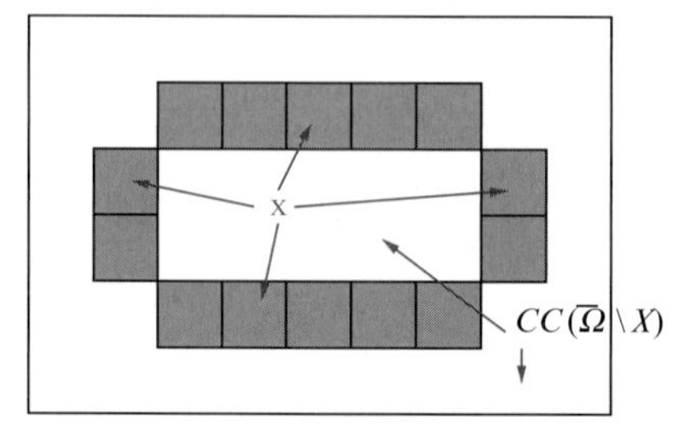

Fig. 6.14 When we consider 8-connectivity for a set and its complement, the gray pixels compose a connected rectangle, but the complement is connected, so that there is no internal hole. This gives a notion of hole not conform to intuition.

On the contrary, when $k = 8$, the answer is, surprisingly, positive. Notice that this gives a counterintuitive notion of a hole, see Fig. 6.14, showing a rectangle having no interior and exterior, that is, the complement is connected. We shall give a hint of the proof.

The increasingness of the saturation operator remains true. For two connected components of level sets of the same type, if they intersect, one contains the other, and thus their saturations are nested in the same order. If they do not intersect, they are disjoint, and each one is included in a hole or the exterior of the other one. The result follows then easily. In the same manner, if we have one connected component of upper level set $C = \mathrm{cc}\,[u \geq \lambda]$ and one of lower level set $C' = \mathrm{cc}\,[u \leq \mu]$, if moreover $C \cap C' = \emptyset$, the same proof applies. Notice that until here, all this stands also for all connections in 4-connectivity. The only remaining case is $C \cap C' \neq 0$. This implies $\mu \leq \lambda$. If $\mathrm{Sat}\, C = \Omega_d$, nothing more is to be proved. Otherwise, the set N of 8-neighbors of $\mathrm{Sat}\, C$ is connected and in $[u < \lambda] \subset [u \leq \mu]$. Only two cases are possible:

1. $N \cap C' = \emptyset$. Since C' is 8-connected and meets $\mathrm{Sat}\, C$, that implies $C' \subset \mathrm{Sat}\, C$, and therefore $\mathrm{Sat}\, C' \subset \mathrm{Sat}\, C$.
2. $N \subset C'$. We get therefore $\mathrm{Sat}\, N \subset \mathrm{Sat}\, C'$. Since $\mathrm{Sat}\, C$ is a connected component of $\Omega_d \setminus N$, we have either $\mathrm{Sat}\, C \subset \mathrm{Sat}\, N$, in which case we get $\mathrm{Sat}\, C \subset \mathrm{Sat}\, C'$, or $\mathrm{Sat}\, N \supset \Omega_d \setminus \mathrm{Sat}\, C$, but since $\mathrm{Sat}\, C \neq \Omega_d$, we have $\mathrm{Sat}\, \Omega_d \setminus \mathrm{Sat}\, C = \Omega_d$, yielding thus $\mathrm{Sat}\, N = \Omega_d$, and therefore $\mathrm{Sat}\, C' = \Omega_d \supset \mathrm{Sat}\, C$.

To allow connection of diagonal neighbors at junction pointels, these junction pointels must have several values. This shows that there is no associated continuous domain image. Instead the discrete image has to be thought as the sampling of a multivalued map. But the theory about an inclusion tree for multivalued images is not done.

The advantage of using only this notion of connectivity is that lower and upper level sets are considered in the same manner. The tree becomes symmetric. Notice however that considering only 8-connectivity prevents the local computation of the Euler characteristic, see Kong and Rosenfeld [49]. The consequence is that the algorithm must be modified in the following way: when a connected component of level set is extracted, to find if it has a hole, we have to devise a standard extraction of the connected components of the complement, see Rosenfeld and Pfaltz [89], Lumia *et al.* [61]. This is an algorithm of linear complexity, but must be done for each connected component of level set found, so that this step becomes the bottleneck of the whole process.

6.4.2 Level Lines Extraction

A top-down algorithm based on level lines is proposed by Song [109]. It extracts first the contour of the image (root level line), then the children and so on. It is based on two key observations: the level λ of a level line is the minimum or maximum of the pixel values at its interior boundary pixels (depending on its type) and its private pixels are those connected either to an interior boundary pixel at level λ or to an exterior boundary pixel at level λ of a child level line.

To fill in the subtree based on the current level line, the principle is thus to start from a pixel on the interior boundary at extremal level λ and do region growing by comparing to the neighbors inside the level line: neighbors at level λ are private pixels and are appended to the pool for region growing; for neighbors at different level, the common edgel is part of a child level line, which is extracted while its exterior boundary pixels at level λ are added to the pool of private pixels.

Apart from minor differences (Song's algorithm is based on a hexagonal grid of pixels and convention for p_∞ is different) the output is equivalent to FLST: level lines vs. level sets, and we can switch from one to the other easily. The complexity of the top-down algorithm is the sum of lengths of the level lines, so optimal since proportional to the output, and is reported in [109] to beat the FLST in execution time on several tests. However this is based on time for FLST + conversion from level sets to level lines: for FLST alone, computation times are similar. It could be noted that the case of multiple bifurcations is not a problem for the top-down approach (can be $O(N)$) while the long branches can be penalizing ($O(N^2)$ if nested shapes have each a single private pixel). One advantage of the level sets approach is that the output is at worst proportional to the number of pixels, which is usually much lower than the lengths of level lines.

A similar method is proposed for the tree of bilinear level lines (Chap. 7), which is made simpler by the fact that level lines are disjoint in this case.

6.4.3 Higher Dimension

The FLST can be generalized to higher dimension, except for one fatal flaw. It concerns the determination of whether a connected component of level set has some internal hole or not: the Euler characteristic is not locally computable in this case, and this ruins the performance of the algorithm. This can be computed as exposed by Lee, Poston and Rosenfeld [55], Lee and Rosenfeld [56], Lumia [60], but not with acceptable complexity.

In 3-D in particular, we would consider 6-connectivity for lower level sets, and 26-connectivity for upper level sets (or the converse). In this case, it is preferable to extract the min- and max-trees and merge them, as exposed in [20]. To the best of our knowledge, there is no efficient "hole-counting" published algorithm. For example, [2] has a different definition of hole, for which the filled torus has one hole, and our notion of hole is referred to as "cavity" and not handled.

Chapter 7
Computation of the Tree of Bilinear Level Lines

Chapter 6 proposed an algorithm computing the tree of shapes based on an interpretation of the discrete image as a piecewise constant bivariate function. The corresponding interpolation operator, commonly known as nearest neighbor or order 0 interpolation, does not removes the pixelization effect. Bilinear or order 1 interpolation is notoriously preferable in this respect. This interpretation of the discrete image as a continuous function by bilinear interpolation is also amenable to a Fast Level Set Transform, which is presented in this chapter. The behavior of level lines of the interpolated function is proven to be easily classified, leading to an algorithm that borrows the same ideas as the FLST with appropriate modifications to handle the singularities. An implementation in the open source software suite MegaWave (http://megawave.cmla.ens-cachan.fr) is in the module `flst_bilinear`.

At this point, the reader may well wonder why more regular interpolations, like higher order spline interpolations (for example bicubic) or even zero padding of the Fourier transform, are not considered. The reason is that the behavior of the level lines becomes much more complex, as they are implicit solutions of higher degree polynomial equations. We are not aware of any work attempting the classification of these behaviors.

7.1 Level Lines and Bilinear Interpolation

Recall that an image can be modeled as a function that assigns a real value (gray level) to each point of a subset of the plane ($S \subseteq \mathbb{R}^2$). A digital image is obtained by sampling the values of this function at discrete positions and by quantizing the gray levels with some finite number of values (typically integer values between 0 and 255). Usually, the sampling of the plane of the image is performed by using a rectangular grid. Each element of this grid is called a *pixel*. For the moment, let us interpret the term level line as the boundary of a connected component of a level set of the image. When we take a close look at the level lines of a digital image (see Fig. 7.1) one important characteristic we observe is the **pixelization** effect. That is, the points which describe the

V. Caselles and P. Monasse, *Geometric Description of Images as Topographic Maps*, Lecture Notes in Mathematics 1984,
DOI 10.1007/978-3-642-04611-7_7,

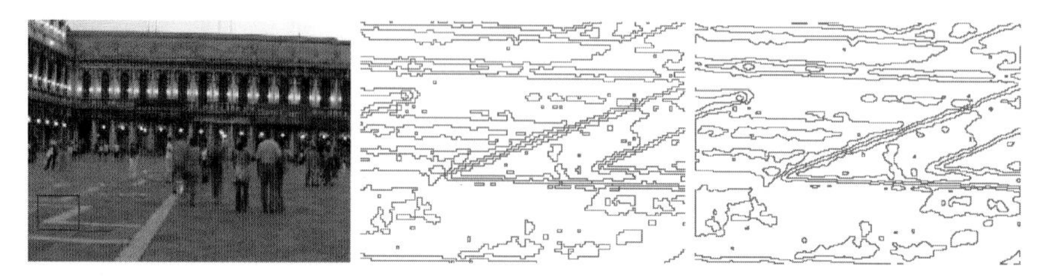

Fig. 7.1 Left: a given image. Right: Pixelized level lines (center) and continuous level lines (right) for the region of top image inside the rectangle (quantization step for the gray levels is 15).

discrete curves take only quantized values, at the vertices of the grid that defines the digital image. Some consequences of this pixelization are:

- level lines are highly irregular, i.e., a lot of sharp edges are present in the curves. As a result, it is difficult to describe the lines by means of characteristic features such as inflexion points or extrema of curvature,
- the accuracy of any measure based on the location of the level lines (e.g., the distance between corresponding level lines in two frames of a video sequence) is limited by the quantization in the position of the curves.
- level lines corresponding to different gray levels may have pieces in common (creating T-junctions). This never happens when dealing with level lines of a continuous image.

A way to get rid of the pixelization effect consists in computing a continuous version of the digital image by interpolation of the discrete values. Once we have this continuous function we can compute its level lines, which has some interesting properties:

- level lines are smoother than in the previous case,
- subpixel accuracy can be achieved when measuring level lines, since they are not restricted to the grid of the digital image,
- since we deal with a continuous function, level lines from different gray levels do not touch, which permits a multiscale visualization.

We use bilinear interpolation, which has the advantage of being the most local (with respect to the values on the grid) of interpolations. Moreover, bilinear interpolation preserves the order between the gray levels of the image and it is in this sense an increasing operation on images.

Let us describe with more detail the bilinear interpolation process. To fix ideas, we assume that the digital image u is given on a rectangular grid $(\mathbb{Z} \times \mathbb{Z}) \cap ([0, M-1] \times [0, N-1])$, $M, N \in N$, and takes values in a finite set F (typically, the integers $\{0, \ldots, 255\}$). We want to compute $\tilde{u} : [0, M-1] \times [0, N-1] \to \mathbb{R}$, the bilinear interpolate of u. The function $\tilde{u}$ is the result of the convolution of u (considered as a network of Dirac masses) with the separable function $\varphi(x)\varphi(y)$, where

$$\varphi(x) = \max(1 - |x|, 0).$$

As $\varphi \geq 0$, bilinear interpolation is an increasing operator. Since $\forall x \neq 0$, $\varphi(x) < \varphi(0) = 1$, the extrema of $\tilde{u}$ are all located at points on the discrete grid. More precisely, all regional extrema of $\tilde{u}$ contain at least a local extremum in the original grid.

The general form of a bilinear function is $f(x, y) = axy + bx + cy + d$. Four equations are needed in order to compute parameters a, b, c and d. If we consider the gray levels of the image to be concentrated at the centers of the pixels, a bilinear interpolate can be computed for each square of four neighbor pixels (which is called a *Qpixel* from now on), see Fig. 7.2.

The continuous function resulting from this interpolation looks like the image in Fig. 7.3. By computing this bilinear interpolation for all the *Qpixels*

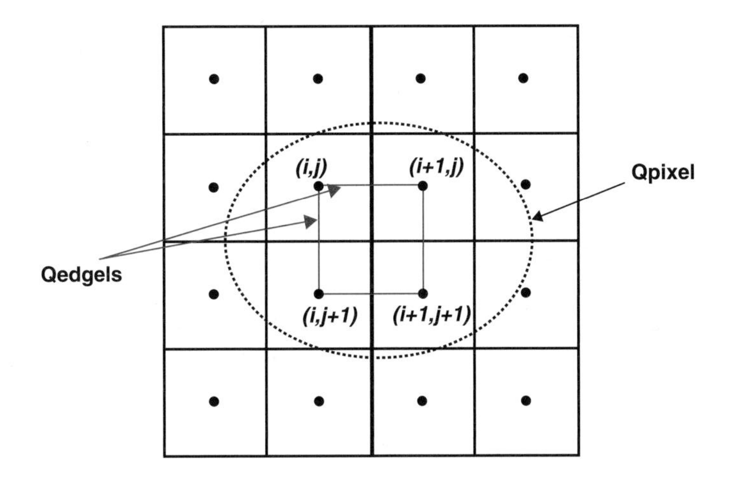

Fig. 7.2 Definition of *Qpixels* and *Qedgels*.

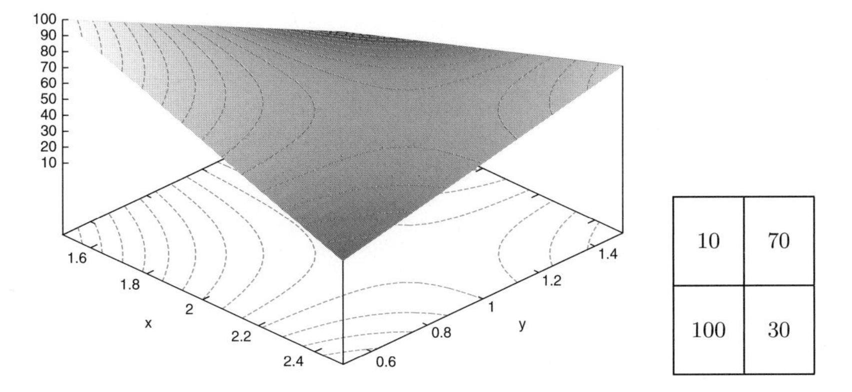

Fig. 7.3 Bilinear interpolation (right) of the gray levels of a *Qpixel* (left).

70	100	70	30	100
70	100	10	100	10
100	70	70	70	30
30	10	70	70	100
70	100	30	30	10

Fig. 7.4 Test image (5×5 pixels) for bilinear interpolation.

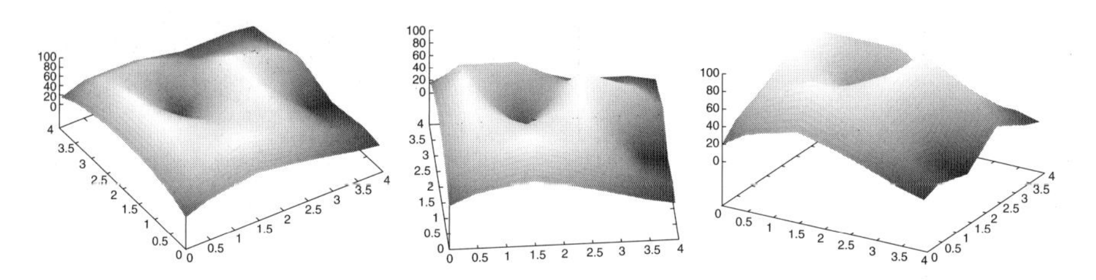

Fig. 7.5 Three views of the piecewise bilinear interpolation of the gray levels of the previous image.

in the digital image, we get a piecewise bilinear function, like, for digital image of Fig. 7.4, the one displayed in Fig. 7.5. Continuity of the gray levels between contiguous "pieces" of this function is guaranteed by the properties of the bilinear interpolation, but higher continuities (e.g., of the gradient) are not preserved at *Qedgels*.

The equation for the level line at level λ of the bilinear interpolate of a *Qpixel* can be written as follows:

$$a(x - x_s)(y - y_s) + (\lambda_s - \lambda) = 0 \tag{7.1}$$

This is the equation of a hyperbola, where $x_s = -\frac{c}{a}$, $y_s = -\frac{b}{a}$ and $\lambda_s = d - \frac{bc}{a}$.

When $a \neq 0$, $\lambda = \lambda_s$ the level line is composed of two perpendicular straight lines that cross at point (x_s, y_s). This singular point where the level line makes a cross is a saddle point and we call saddle level the gray level λ_s for which we have such a situation. Every bilinear interpolation of a *Qpixel* has an associated saddle point (the center of symmetry of the hyperbola), but it is not always inside the *Qpixel*.

Figure 7.6 displays some of the level lines for the bilinear interpolate of a *Qpixel*.

Definition 7.1. We call level line any connected component of $[\tilde{u} = \lambda]$ for any $\lambda \in \mathbb{R}$.

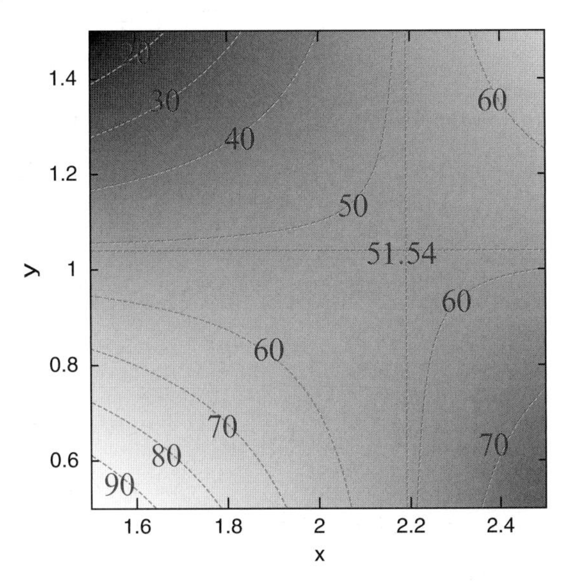

Fig. 7.6 Level lines (pieces of hyperbolae) for the interpolated *Qpixel* in Fig. 7.3. Displayed level lines are from level 20 to 90 with quantization step 10. There is a saddle point for gray level $10 + 60 \times 9/13 = 51.54$.

Observe that, by the above discussion, level lines for the piecewise bilinear interpolation of a digital image associated to values $\lambda \notin F$ will be a concatenation of pieces of hyperbolae. Some remarks can be made concerning these level lines. Let us define a *Qedgel* as the segment joining two centers of adjacent pixels (Fig. 7.2). If none of the levels of the level lines is equal to the original levels of the image, we can guarantee that the level lines will never cross the centers of the pixels nor will they follow the *Qedgels*, but always cross them. Choosing the gray levels this way guarantees that no more pixelization effect is present in the level lines. Moreover, a level line will cross a *Qedgel* only once (at most). The sequence of all the crossings of a level line with the *Qedgels*, together with its saddle points, is the minimal discretized description of a curve coming from bilinear interpolation. Some of the level lines for image in Fig. 7.5 are displayed in Fig. 7.7.

In order to distinguish between the level lines corresponding to values $\lambda \notin F$ and the values $\lambda \in F$ we shall call them *bilinear level lines* and *classical level lines* respectively.

7.2 Tree of Bilinear Level Lines

In Chaps. 2 and 6 the tree of level sets was defined and computed by the FLST algorithm ([79]). The extraction of this tree was carried out by considering the level sets of the digital image. The information on the inclusion relations between the connected components of level sets of the image (the *shapes*) was coded in this tree, in such a way that a *shape* is *child* of another *shape* if it

Fig. 7.7 Level lines of the piecewise bilinear interpolated image in Fig. 7.5. Three sets of level lines are displayed. Left: levels 10 to 100 with step 10. Observe how some of the level lines (at gray level 70) follow the *Qedgels*, producing an effect similar to pixelization. Center: levels 11 to 91 with step 10. Pixelization effect no longer arises since level lines are computed at gray levels different to those of the original image. Right: level lines at gray levels different from those of the original image. There are 90° crossings of level line due to the presence of a saddle point. Nevertheless, these saddle points always appear inside the *Qpixels* and the curves never go along the grid of the digital image.

is included in its *interior*. The notion of interior depends on the type of level set (upper and lower) and a different connectivity (4 or 8) needs to be used in order to extract the interior of an upper or a lower level set.

The framework was based on semicontinuous images. When we deal with continuous images, the property that level lines are disjoint provides a much simpler argument for the existence of the tree structure, similar to that provided in the pioneering work of Kronrod [50]. The tree structure is however different from that of [23] where the order in the tree is driven by the gray level and not by a geometric consideration.

When bilinear level lines are used, since they never touch, a fast and simple algorithm can be devised, based on the crossings of the level lines with the *Qedgels* of the image.

In Chap. 2, we avoided defining level lines but used rather level sets, as they are more natural for semicontinuous images. Now in a traditional topographic map of terrain we are accustomed to seeing lines of isoelevation. As these lines are disjoint, visualization is facilitated. We give a first definition of level line. We will call level lines the boundaries of the connected components of the upper level sets $[u \geq \lambda]$:

$$\{\mathcal{CC}(\partial[u \geq \lambda]) : \lambda \in \mathbb{R}\}\,.$$

In most cases, these level lines are what we expect: Jordan curves at constant elevation λ. However, it is readily apparent that these do not account for regional minima of the image. Also at saddle points the level lines are not Jordan curves anymore. We see here the limitations of this definition and will later refine it. For the moment, we assume that the level lines are simple

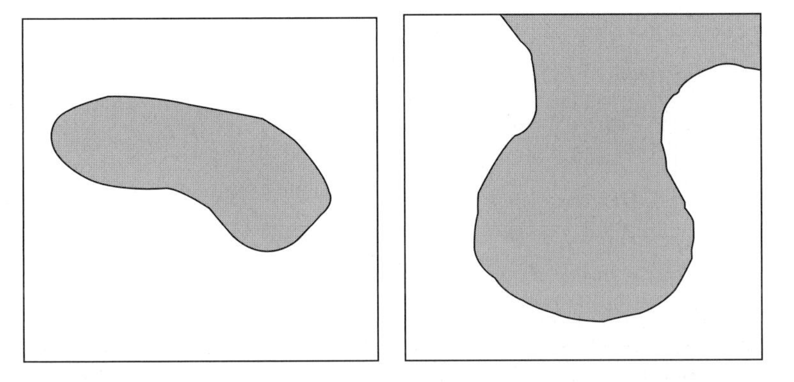

Fig. 7.8 Interior (in gray) of a closed and an open level line. Open level lines are closed following the shortest path along the border of the image.

curves and we order them by inclusion in a tree. For this, we define the interior of a level line as shown in Fig. 7.8: for a closed level line it is the inner domain and for an open level line (one meeting the boundary of the image), it is the inner domain of the closure of the line by the shortest path along the image boundary. Notice that this definition of interior is not based on whether the level is higher in the interior or not, but on geometry.

From this definition it is natural to define a partial order on these level lines by comparing their interiors, which are either nested or disjoint.

Observe that the bilinear interpolate of the data points, $\tilde{u}$, is continuous in $\overline{\Omega}$ and has a finite number of regional extrema, i.e., it is a weakly oscillating function. By Lemma 4.12 and Proposition 4.23 of Chap. 4 we have the following result (which also admits a simpler proof):

Proposition 7.2. *a) At any level $\lambda \in \mathbb{R}$, the number of level lines at level λ is finite.*

b) If L is a level line at level $\lambda \in \mathbb{R}$, the image $\tilde{u}$, in the exterior vicinity of L, is either uniformly $< \lambda$ or $> \lambda$.

In particular, this implies that the interior of a level line contains a local extremum, a fact that will be exploited in our second algorithm.

7.3 Algorithms for the Extraction of the Tree of Bilinear Level Lines (TBLL)

We propose two algorithms for the extraction of the TBLL of an image. The first one is simpler but requires a quantization avoiding the initial levels of the image. The second one is a variant of the FLST of Chap. 6, computing the so called *fundamental* TBLL of the image, describing the topography of the image, which in turn can be used to compute the TBLL corresponding to *any* quantization.

The data structure for the level lines can be the same as for shapes in Chap. 6, except that we add a list of points coding the sampled level line to each shape.

7.3.1 Direct Algorithm

This algorithm is in two phases: first extract the bilinear level lines, then order them in a tree.

The extraction of a level line at $\lambda \notin F$ relies on these observations:

1. the level line is *not* a regional extremum, so that an orientation can be chosen, for example leaving the upper level at the left of the level line;
2. the level line meets *Qedgels* at non endpoints, so that when we "enter" a *Qpixel* by a *Qedgel*, we can compute from which other *Qedgel* to "exit".

Considering that we get into the *Qpixel* through some *Qedgel*, the exit *Qedgel* is in front, on the left or on the right depending on the value of λ with respect to the levels at both other corners of the *Qpixel*, see Fig. 7.9. Notice that when $\lambda = \lambda_s$, we exit on the right hand side. We may store for the line only its list of intersections with *Qedgels*, or store intermediate points by using (7.1).

To avoid extracting several times the same level line, we put markers at all intersection points of extracted level lines with *Qedgels*. The algorithm starts by considering all *Qedgels* at the boundary of the image. If a level of interest λ is between the levels at endpoints of the *Qedgel*, we follow the level line, marking intersection points with *Qedgels*. We then close the level line by the shortest path along the image boundary. Finally we do the same for interior *Qedgels*. For them, we detect closure by checking the marker at exit *Qedgels*.

We refer to Algorithm 7 to build the tree structure of the extracted level lines. It considers all intersection points of level lines with vertical edgels and orders them. While scanning a column of *Qedgels*, we are in the interior domain of a level line if we have crossed it an odd number of times. If we meet a level line that has no parent yet, its parent has to be P, the last level line we are in. For the root, we add the boundary of the image as a level line to the list L at the beginning.

An example of the inclusion tree computed from bilinear level lines is shown in Fig. 7.10.

This algorithm relies on the fact that the quantization is chosen so that each level line crosses *Qedgels*, but does not contain any. For this, it is sufficient that it avoids the levels at the centers of pixels (the initial data). In particular, regional extrema of the image cannot be extracted by the algorithm. On the contrary, the second algorithm we propose deals explicitly

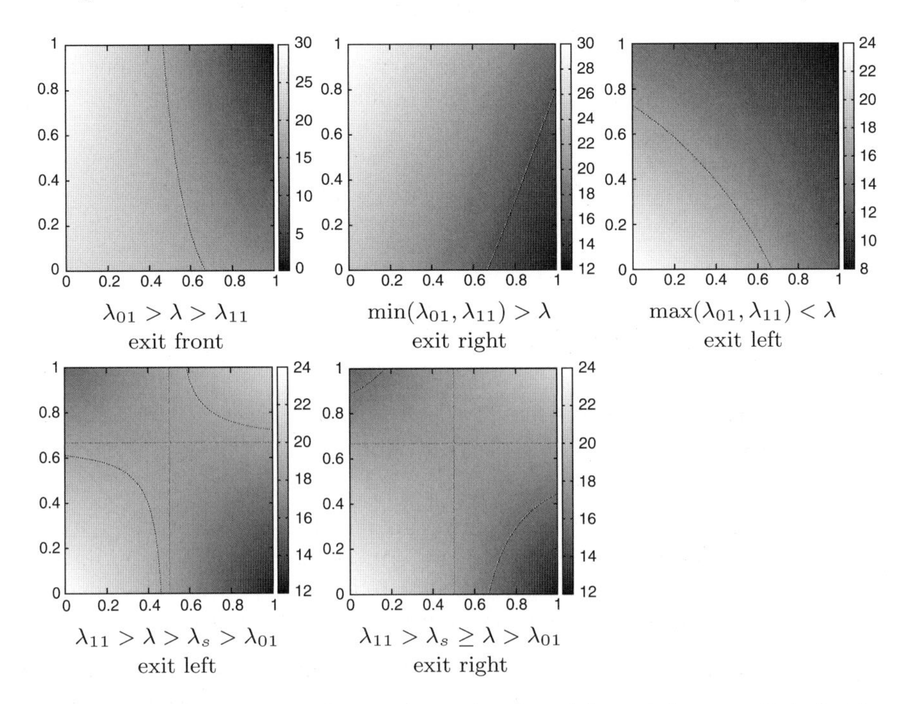

Fig. 7.9 Direction to exit when entering the *Qpixel* through bottom *Qedgel* while following a level line at λ. By convention, $\lambda_{0,0} > \lambda > \lambda_{0,1}$. The *Qedgel* through which to exit depends on the position of λ compared to the other two levels involved, λ_{01} and λ_{11}. In the two bottom cases, we have a saddle point inside the *Qpixel* at level λ_s.

Input: List of bilinear level lines L
Output: Fill tree structure of L
Collect all (i, y, id) in array V `/* Intersection of curve index` id `with` *Qedgel* `at column` $i + 0.5$ `*/`
Order V lexicographically by key (i, y)
$P \leftarrow \emptyset$ `/* Innermost shape */`
for $(i, y, id) \in V$ **do**
  if $P = L_{id}$ **then** $P \leftarrow P$.Parent `/* Getting out of innermost shape */`
  else
    if $P \neq \emptyset$ **and** $L_{id}.Parent = \emptyset$ **then** Set L_{id} as child of P
    $P \leftarrow L_{id}$ `/* Innermost shape is now` L_{id} `*/`

Algorithm 7: Ordering the level lines in a tree.

with them. Another problem is related to saddle points. As is apparent in Fig. 7.10, B has a loop around C but there is another loop around Q, which is a separate level line child of B. This behavior is dependent:

- on the orientation of B, counterclockwise here, because we assumed that levels at the left are higher;
- on the convention to turn right when arriving at a saddle point.

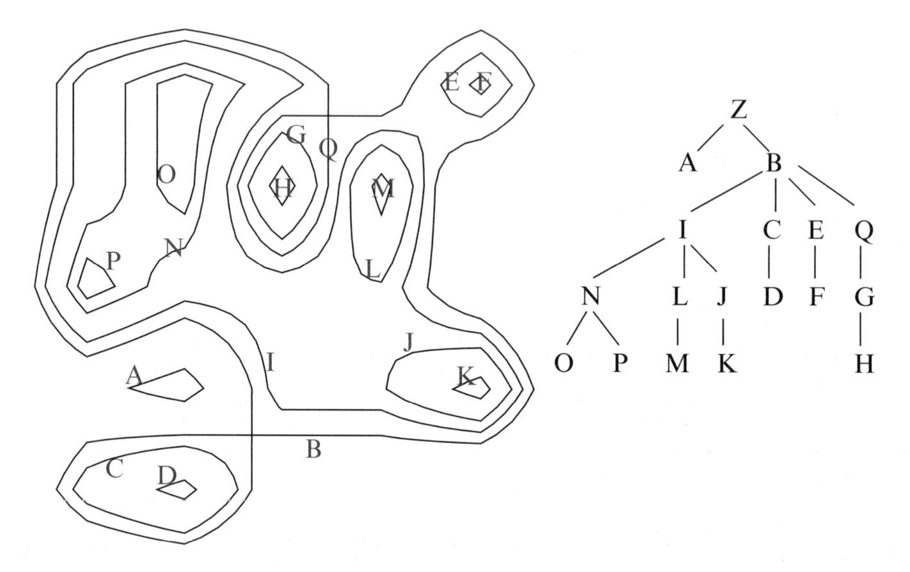

Fig. 7.10 Inclusion tree (right) from bilinear level lines (left).

This is the same behavior we would observe for the Morse algorithm presented below, except that Q would not be considered a level line. However, for the inverse image, the orientation of B would be inverted and the loop around C would be a new level line, sibling of A and B, while Q would be part of B and G another child of Z. We see that this interpretation of the same topographic map is not self-dual, another motivation to deal explicitly with singular points, as the Morse algorithm does.

7.3.2 Morse Algorithm

We call so our second algorithm as it extracts level lines at critical points, that is, extrema and saddle points (and also at data points). As for a Morse function, these levels correspond to a change in the topology of level lines.

We call fundamental TBLL of the image the tree of level lines passing through a center of pixel or a saddle point. Therefore, all level lines containing critical points are in the fundamental TBLL. All other level lines can be deduced from those.

We give here a more appropriate definition of level line than the one proposed above: we will call level line the boundary of a shape. This definition has the advantage of having the ability to represent all levels of the image. It is however more involved because it defines the interior (the shape) before the level line (see Fig. 7.11). A level line is almost a Jordan curve: it can have (a finite number of) double points, occurring at saddle points.

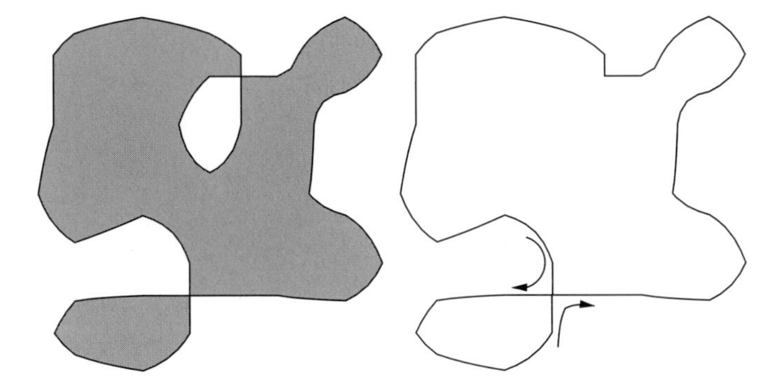

Fig. 7.11 Correct definition of a level line. Left: a connected component of level set containing two saddle points. Right: the associated level line, which can be oriented as shown.

We will say that a *Qpixel* or *Qedgel* P is adjacent to the shape S if

$$P \setminus S \neq \emptyset \neq P \cap S;$$

in particular, P meets the boundary of S. Since S and its complement are connected, the *Qpixels* adjacent to S can be ordered in a chain, so that each node is 4-adjacent to the following one (two *Qpixels* are 4-adjacent if they share a common *Qedgel*). As two successive *Qpixels* in this chain share a *Qedgel* adjacent to S, we can store either the chain of adjacent *Qpixels* or of *Qedgels*.

Proposition 7.3. *A Qedgel E is adjacent to a shape S if and only if exactly one of both endpoints of E is in S. $E \subseteq S$ if and only if both endpoints of E are in S. $E \cap S = \emptyset$ if and only if both endpoints of E are not in S.*

These properties can be easily proved using Proposition 7.2.b and the fact that the restriction of the image to E is affine.

A consequence is that the datum of the chain of adjacent *Qpixels* to a shape S is equivalent to the knowledge of the centers of pixels in S (see Fig. 7.12).

The main reason for computing the fundamental TBLL is the following set of properties:

1. For each level line L'' in the TBLL, there are some shapes L and L' in the fundamental TBLL such that L is the parent of L' and L'' is comprised between L and L'.
2. If L and L' are in the fundamental TBLL, L' being a child of L, and if L'' is any level line comprised between L and L' in the TBLL, the chains of *Qpixels* adjacent to the shapes associated to L' and L'' are the same.

The second property comes from Proposition 7.3: indeed, the shapes associated to L' and L'' contain the same centers of pixels, implying that their

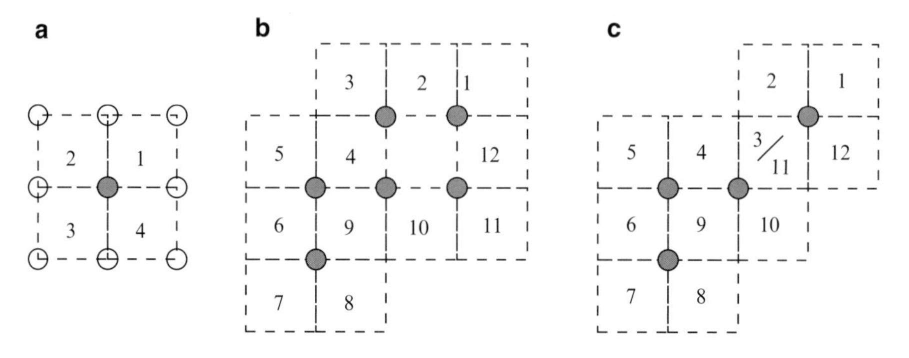

Fig. 7.12 Chains of *Qpixels* adjacent to shapes. Centers of pixels in shape are represented as gray dots. **a**: only one center of pixel in the shape. **b**: a standard configuration. **c**: 8-connection is made by the intermediary of a saddle point, the *Qpixel* containing it is present twice in the chain, at positions 3 and 11.

adjacent *Qedgels*, and thus *Qpixels*, are the same. This property shows that the knowledge of the chains of 4-adjacent *Qpixels* of level lines of the fundamental TBLL permits to deduce the ones in any TBLL.

Extraction of the Fundamental TBLL The algorithm of extraction of the fundamental TBLL is almost identical to the one exposed in Chap. 6, except for the changes we outline here. It follows from Proposition 7.2, extracting interiors of level lines rather than directly the level lines themselves.

We first find the *Qpixels* containing a saddle point, and the corresponding saddle value. It is the case when both values at diagonally opposed corners are strictly less than the other two values. In this algorithm a *point* is either a center of pixel (i.e., a data point) or a saddle point. Each point has an associated value, so we store in memory the values at centers of pixels and the values at saddle points. In the implementation code, it may be more convenient to have a separate array for values of saddle points, with a special value meaning the absence of a saddle point in the *Qpixel*. Place holders for saddle points must also be inserted in the image of flags `ImageNeighbors` of page 121. Again, a separate image of flags may be used for saddle points.

The saddle points modify the adjacency graph according to Fig. 7.13. We see that an interior data point can have between 4 and 8 neighbors (5 in Fig. 7.13 right), while a saddle point has always 4 neighbors.

In Algorithm 2, the test of local extremality is done according to this new adjacency graph. However, since saddle points have intermediate values, it is equivalent to comparison with the 4 adjacent data points (4-connectivity); moreover, saddle points need not be tested for extremality.

In the same manner, the modification of the Euler characteristic when adding a point to a region must be adapted to the new graph. The patterns of Figs. 6.7 and 6.8 must be used *both* for a data point, with the understanding that a diagonal neighbor either means the corresponding saddle point if present, or is considered as an outside point otherwise. It can be readily

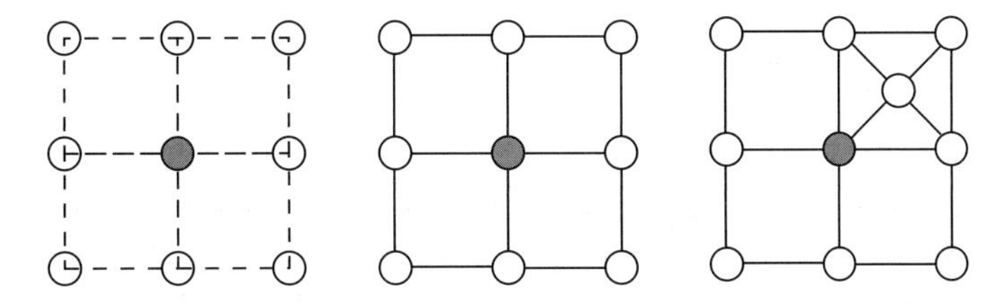

Fig. 7.13 Modification of the adjacency graph in the presence of a saddle point. Left: a data point and the 4 associated *Qpixels*. Middle: adjacency graph in the absence of saddle point. Right: adjacency graph in the presence of a saddle point in the upper-right *Qpixel*.

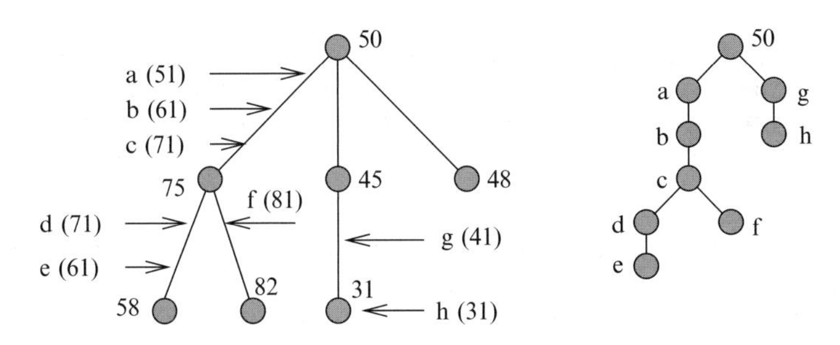

Fig. 7.14 Computing the TBLL of quantization levels $\{1, 11, 21, \ldots\}$ from the fundamental TBLL. Left: fundamental TBLL, showing associated levels. Right: the resulting TBLL is obtained by sampling of the fundamental TBLL.

verified that when a configuration can be represented by patterns in both Fig. 6.7 and Fig. 6.8, the variation of the Euler characteristic is the same. Concerning the addition of a saddle point to the set, the only case it modifies the Euler characteristic is when exactly two diagonal neighbors are in the set, in which case a new hole is created.

Computation of a TBLL from the Fundamental TBLL From the knowledge of the fundamental TBLL and a given quantization, it is direct to compute the resulting TBLL. For each shape S in the fundamental TBLL, find the levels of the quantization comprised between the level of S (strictly) and the level of its parent S' in the fundamental TBLL. Create a shape for each one; the level line passes through the same chain of adjacent *Qpixels* as S'.

For each shape created in this manner, their order in the TBLL is the same as the one of their least greater shape in the fundamental TBLL. The TBLL can be considered a sampling of the fundamental TBLL, see Fig. 7.14. To recover the level line itself, and for example store its polygonal sampling, we can get its list of adjacent *Qedgels* from the chain of adjacent *Qpixels*. To add intermediate points, we can use (7.1) to parameterize the level line inside a *Qpixel* as either $x = f(y)$ or $y = g(x)$ depending on which of the components of the gradient is greater.

Chapter 8
Applications

We gather here some applications of the FLST and the tree decomposition to image processing. All were published in the literature, and although they do not pretend to represent the state of the art in their respective domain, we hope they emphasize its versatility and convenience. In several instances, level lines provide a superior alternative for detection based on edges because of their linked structure and indepedence from parameters.

8.1 Image Filters

Apart from self-dual grain filters, as exposed in Chap. 3, other kinds of self-dual morphological operators can be designed based on the tree of shapes. Keshet uses it to define self-dual erosions [48]. In mathematical morphology, erosions are defined by a structuring element B (a set of pixels), where each pixel P gets the minimum of pixel values in the set $P+B$, i.e., the structuring element translated to P. Dilations are defined similarly by the maximum in $P+B$. An opening is an erosion followed by a dilation.

A self-dual erosion is a kind of erosion on the unfolding transform of the image. The unfolding transform replaces each pixel by the depth of its smallest containing shape in the tree. The erosion amounts to setting the value of P to the level of the smallest shape containing all pixels of $P+B$. Self-dual dilation cannot be defined in general, but only on an image resulting from erosion. This puts in P the level of the largest shape containing all pixels in $P+B$, which exists if the image was eroded by B. Their composition is the self-dual opening.

Figure 8.1 shows the self-dual erosion and opening on an image of a Kandinsky painting. Notice that the dark ring and the lighter disk in the upper-left both shrink in self-dual erosion, an effect impossible to get with any classical erosion.

V. Caselles and P. Monasse, *Geometric Description of Images as Topographic Maps*, Lecture Notes in Mathematics 1984,
DOI 10.1007/978-3-642-04611-7_8,

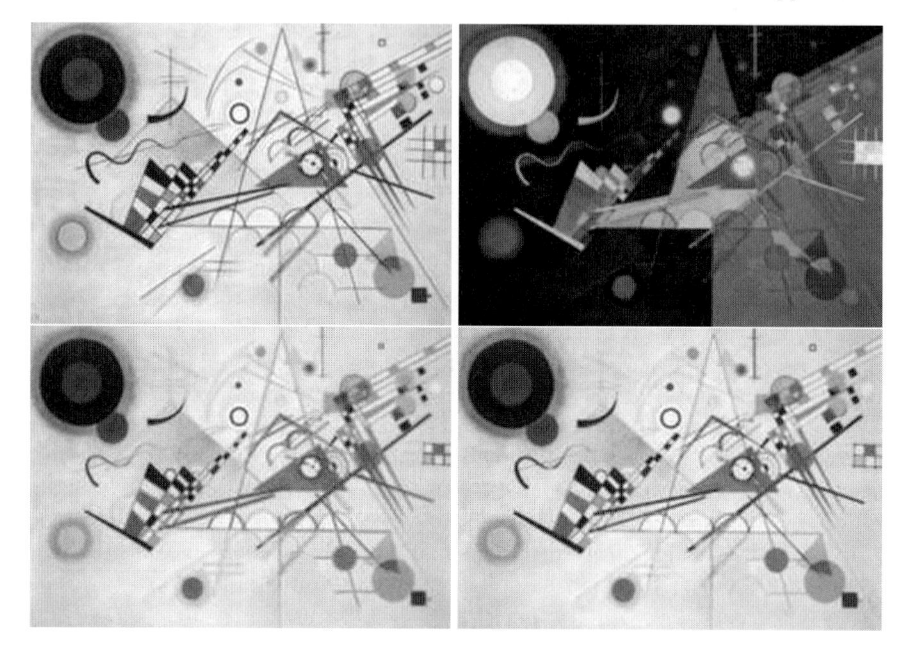

Fig. 8.1 Self-dual morphological operators. Top row: original image and its unfolding transform. Bottom row: its self-dual erosion by a cross structuring element and its self-dual opening.

8.2 Image Registration

Aligning two images that partially overlap seems to be a trivial task compared to optical flow or disparity estimation because it consists in the estimation of a much smaller number of parameters. These parameters can only account for specific phenomena, like observation of a planar scene or pure rotation of a camera, sometimes some elastic deformation as in medical imaging. The main difference comes from the fact that registration deals with global or large scale changes while the latter ones deal with local motion. The precision of the registration is crucial to the quality of downstream treatments, like mosaic or panorama generation, super-resolution or change detection.

The traditional technique consists in minimizing a distance between the two images (considered as functions) among the parameter space. A favorite measure is the L^2 norm of the difference, as it translates to maximal cross-correlation, which can be computed efficiently through the Fast Fourier Transform. In many ways, these metrics fail in several of the desirable qualities of a robust registration method:

- insensitivity to outliers (estimate a dominant motion of precision unaffected by local displacements);
- insensitivity to global illumination variations;

- ability to handle small overlap;
- efficiency for a large parameter space.

The last point means the ability to directly estimate many parameters without simulation. For example, correlation handles efficiently only the translation, other parameters (like rotation, zoom) must be handled by sampling their possible values and simulating the motion on one image. If the sampling must be fine and several parameters need simulation, the computational burden becomes overwhelming.

Another approach relies on invariant descriptors of local interest regions. SIFT (Scale Invariant Feature Transform) neighborhoods and descriptors satisfy similarity and illumination invariance while being fast to compute [59]. Level lines have also these qualities, as demonstrated in [77]. A subset of these, called MSER (Maximally Stable Extremal Regions), were independently rediscovered and used for stereo disparity estimation and have since become very popular [68]. A comparison between variants of SIFT and MSER concluded of the superiority of MSER over SIFT in all cases except in case of blur or strong JPEG compression [74]. It can be argued that these exceptions are due to the scale simulation of SIFT (by properly smoothing and scaling down the images, these defects are reduced).

The approach presented here does not worry much about false matches as long as true ones are detected. The number of outliers may be allowed to be large because they do not participate in the maximal consensus. Therefore the number of matches will be more important than their reliability and we can afford to consider all level sets of sufficient area, not just contrasted ones.

8.2.1 Shapes as Features and Their Descriptors

We use saturated connected components of level sets (output of FLST) as features. We ignore those meeting the image boundary as they are likely partly occluded. We encode them by combinations of moments that are similarity invariant. We note the moments m_{ij} of a shape S

$$m_{ij} = \iint_S x^i y^j \, dx dy$$

and the centered moments μ_{ij}:

$$\mu_{ij} = \iint_S (x - \frac{m_{10}}{m_{00}})^i \, (y - \frac{m_{01}}{m_{00}})^j \, dx dy.$$

The $\mu_{i\,j}$ are translation invariant. The inertia matrix, defined by

$$I_S = \begin{pmatrix} \mu_{2\,0} & m_{1\,1} \\ m_{1\,1} & \mu_{0\,2} \end{pmatrix}$$

is transformed by a rotation R (centered at centroid of shape) to $I_{RS} = RI_S R^T$, so that the trace and determinant of I_S are rotation invariant. The moments of the zoomed shape λS are related to the moments of S by the relation $\mu_{i\,j}(\lambda S) = \lambda^{i+j+2}\mu_{i\,j}(S)$. From moments of order $i + j \leq 2$ we can build 2 similarity invariants:

$$s_1 = (\mu_{2\,0} + \mu_{0\,2})/m_{0\,0}^2 \quad s_2 = (\mu_{2\,0}\mu_{0\,2} - \mu_{1\,1}^2)/m_{0\,0}^4.$$

Two invariants based on order 3 moments can be added:

$$s_3 = \frac{(\mu_{3\,0} - 3\mu_{1\,2})^2 + (\mu_{0\,3} - 3\mu_{2\,1})^2}{m_{0\,0}^5} \quad s_4 = \frac{(\mu_{3\,0} + \mu_{1\,2})^2 + (\mu_{0\,3} + \mu_{2\,1})^2}{m_{0\,0}^5}.$$

However higher order moments are more sensitive to noise so we cannot go to far in the number of descriptors based on moments.

Notice that moments are easy to compute in linear complexity in number of pixels by using the inclusion tree structure. More elaborate descriptors can be adopted, for example based on histograms of gradient directions inside the shape, as in SIFT descriptors [59], but good results can be obtained with the above simpler descriptors.

8.2.2 Meaningful Matches

Two shapes match according to the closeness of their invariant descriptors s_i. Rather than fixing thresholds for index i, we use an *a contrario* approach (see [26] for a full account of the theory), assuming that the descriptors are independent. We would say that shape S (in left image) matches to S' (in right image) if a random shape in right image matches better to S or a random shape in left image matches better to S'. We use an empirical model for a random shape, or rather for its descriptors. We assume that the descriptor s_i of a random shape follows the probability density function φ_i, obtained by normalizing the histogram of observed s_i. Moreover, we assume that these random variables are independent. The probability that S matches better to S' than a random shape in left image is:

$$P(S \to S') = \prod_i \int_{s'_i - |s'_i - s_i|}^{s'_i + |s'_i - s_i|} \varphi_i(x)\,dx.$$

The probability that S' matches better to S than a random shape in the right image is:

$$P(S \leftarrow S') = \prod_i \int_{s_i - |s'_i - s_i|}^{s_i + |s'_i - s_i|} \varphi'_i(x)\, dx,$$

the φ'_i representing the empirical models for the descriptors in the right image. The two events are highly correlated, so we define the probability of the match of S and S' as

$$P(S \leftrightarrow S') = \max(P(S \rightarrow S'), P(S \leftarrow S')).$$

Notice that in the discrete setting, the integral

$$\int_{s'_i - |s'_i - s_i|}^{s'_i + |s'_i - s_i|} \varphi_i(x)\, dx$$

becomes just the proportion of shapes in left image whose descriptor i is closer to s'_i than s_i is.

We define the number of false alarms of the match as the probability multiplied by the number $N_l N_r$ of possible pairs of shapes, N_l (resp. N_r) being the number of shapes in left (resp. right) image.

$$NFA(S \leftrightarrow S') = N_l N_r P(S \leftrightarrow S').$$

The match is said to be meaningful if $NFA(S \leftrightarrow S') \leq 1$.

8.2.3 Grouping Matches

Monotone sections group shapes by inclusion. Many of these shapes are similar, and it is advisable to keep them grouped for matching. This reduces the number of matches to consider, as we would consider matches of monotone sections instead of matches of individual shapes. To still indicate that some sections are more important than others, based on their number of shapes, an integral weight is attributed to matches of sections. The weight of the match has to be linked to the number of matches between shapes of the sections. But we count only compatible matches of shapes; two such matches are said compatible if they preserve strict inclusion. This is illustrated as a graph in Fig. 8.2. Notice that this also enforces uniqueness: if a shape matches two different shapes in the other section, these matches are incompatible because we have strict inclusion on one side not preserved on the other side. We consider the weight of the match of sections the cardinal of a maximal subset of matches of shapes that are two-by-two compatible. Notice that the weight is at most the minimal number of shapes in either section.

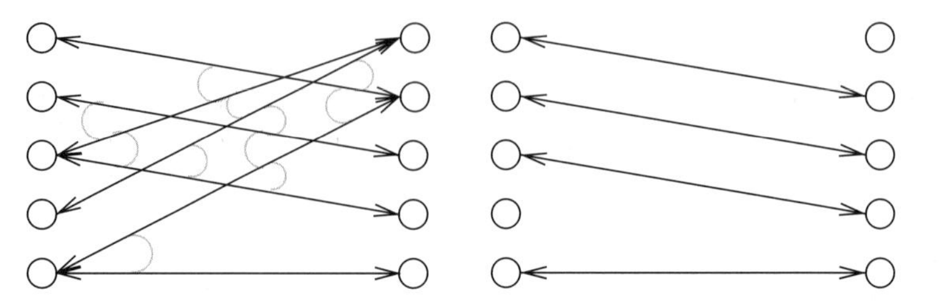

Fig. 8.2 Compatibility of matches of shapes between monotone sections. Left: Incompatible matches of shapes. Left nodes represent shapes in one monotone section and right nodes in another monotone section. They are ordered by inclusion from top to bottom. All intersecting edges represent incompatible matches. Right: A maximal subset of compatible matches. The match of sections has weight 4.

The similarity with the ordering constraint along scanlines in stereo disparity computation is obvious [86]. The weight of the match is also efficiently computed by a dynamic programming algorithm.

8.2.4 Eliminating Outliers by Vote

Notice that in general, whatever the robustness of the local descriptors, false matches are bound to happen. This is because in artificial environments repeated structures are frequent, like windows on a building facade, tiles on a floor or rooftops in aerial images. A consequence is that any minimization procedure for the estimation of the parameters that does not account for outliers will fail in giving highly accurate results. The solution adopted in the SIFT method is to ignore a match that is not significantly better than any other match sharing a common shape. In other words, SIFT ignores repeated structures, expecting to still have enough unique features to match. In scenes with many repetitions, as illustrated later, the number of accepted matches may then become insufficient.

For complexity reasons, and because of the limited geometrical invariance of the descriptors, we restrict to the case where the transform mapping the two images is correctly approximated by a similarity. We have thus 4 parameters to estimate, the rotation angle, the zoom factor and the two translation components. Whereas the rotation and zoom commute, the translation does not commute with them. That means that they are dependent, and we could either try to estimate all together, or in two 2-parameter estimations, which should not be independent. The simplest is to first estimate rotation and zoom, then translation conditionnally to these estimations.

As a linear transform, the similarity transform commutes with centroid extraction, that is, the centroid of the original shape maps to the centroid of the matching shape. From two correspondences $S_i \leftrightarrow S'_i$ and $S_j \leftrightarrow S'_j$, the centroids permit to vote for the 4 parameters. As already written, we vote only for rotation angle θ and zoom factor λ in a two dimensional histogram with a weight $w_i w_j$. The peak of this histogram is selected, and a vote for the two components of translations is done in a two dimensional histogram after application of the zoom and rotation to centroids in the left image. The center of this initial transform is the left image center.

When the translation peak has been found, the matches having voted for the elected similarity are considered inliers and the others outliers. Based on these inliers, a least squares error minimization of the homography transform can be performed.

Notice that this procedure of robust estimation is very close to RANSAC (Random Sample Consensus) of [32], where minimum samples to estimate the model are randomly chosen and the compatibility of other matches is measured. There are two minor differences: we consider all pairs of matches instead of taking random samples, as we can afford it in these examples; and the errors w.r.t. the model are measured in parameter space rather than in image space for RANSAC.

8.2.5 Examples

Results of panoramas (mosaics) shown in this section are voluntarily crude, obtained by averaging overlapping pixels, without any contrast adjustment to make them look more natural, such as multi-band blending [15]. This allows better visual inspection of the geometric precision.

An example of similarity registration is presented in Fig. 8.3. The watch undergoes a rotation and the time changes. In the mosaic, hands of the watch appear as ghosts because their position has changed.

Figure 8.4 shows the registration of two satellite images. The presence of sharp peaks in the histograms indicates no motion ambiguity. The final motion is a true homography, not a similarity. This homography aligns exactly the images.

In Fig. 8.5 we show two images obtained from a camera rotating around its optical center. Notice the motion of the bus in the foreground between the two snapshots. Parts of the histograms of votes for zoom/rotation and translation are shown in Fig. 8.6. The resulting mosaic (Fig. 8.7) shows that the precision is unaffected by the moving bus. The defect in the alignement of line markings may be partially due to geometric distortion of the lens (visible in left image of Fig. 8.5, where the lines are curved instead of straight).

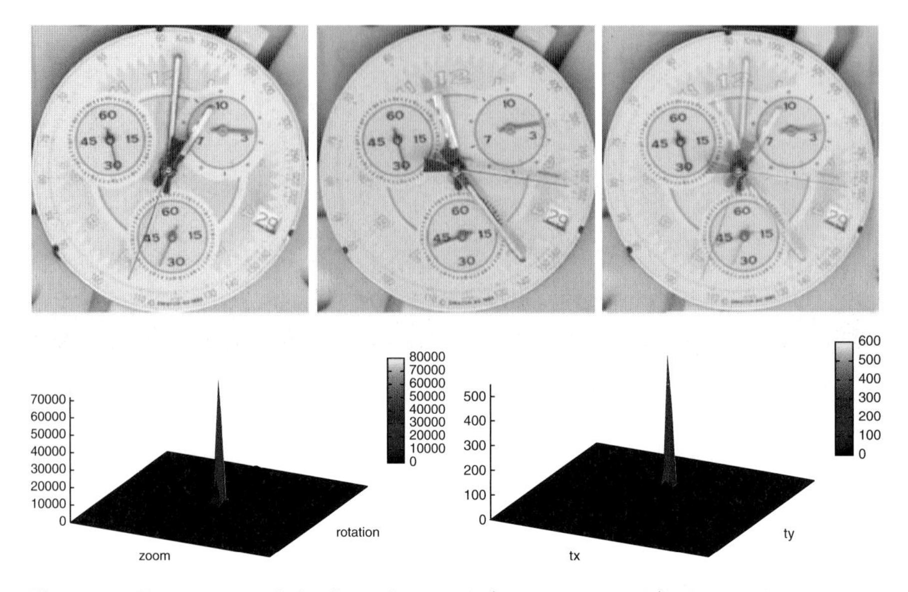

Fig. 8.3 Two images of the face of a watch (512×512 pixels), the resulting mosaic and parts of the histograms of votes.

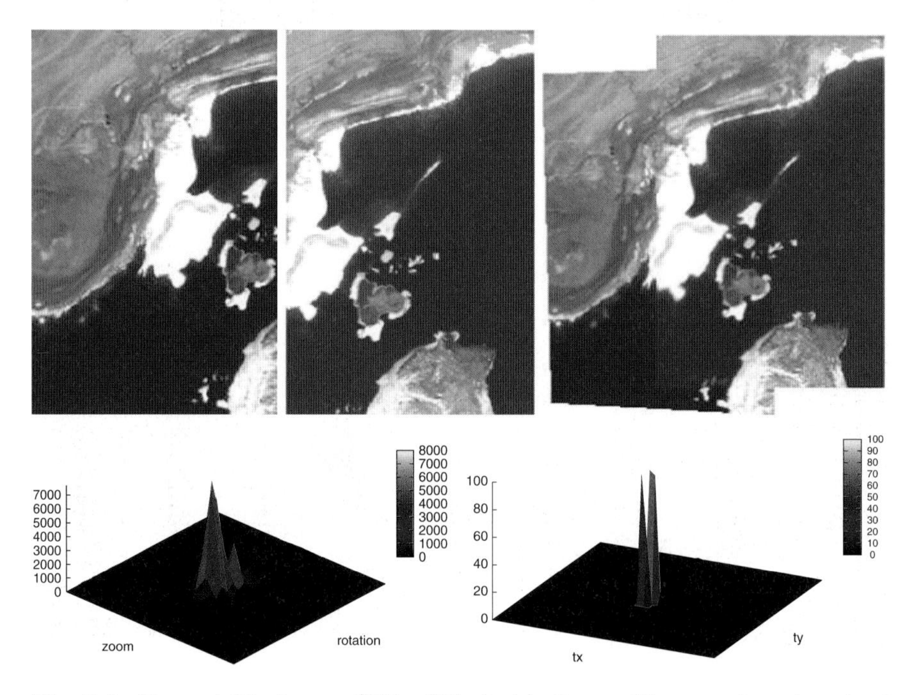

Fig. 8.4 Two satellite images (166×250 pixels), the resulting mosaic and parts of the histograms of votes.

Fig. 8.5 Two pictures (640 × 480 pixels) taken from a rotating camera, used for registration.

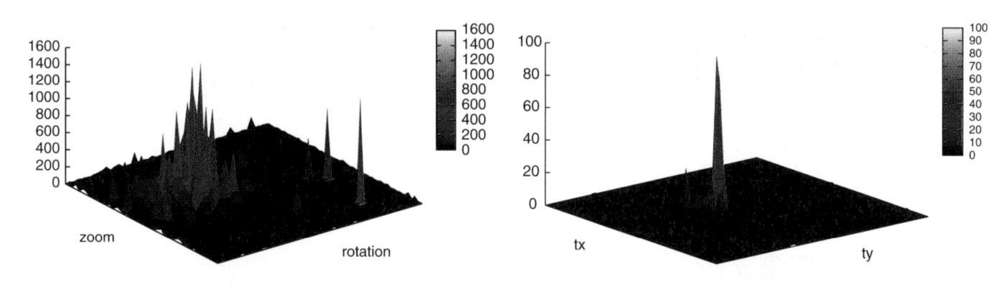

Fig. 8.6 Votes for zoom/rotation and translation in the registration of images of Fig. 8.5.

Fig. 8.7 Mosaic from images of Fig. 8.5. The shadow of the bus is not a registration error but shows that the bus has moved between the two snapshots.

8.3 Other Applications

8.3.1 Image Intersection

As a by-product of the registration presented above, the shapes voting for the elected motion can be used to reconstruct the image, all others being put to the level of the root of the tree. In other words, this process shows the shapes of one image that we can find in the second image modulo the global dominant motion (Fig. 8.8). This can be called the intersection of the images and generalizes the method proposed in [12].

8.3.2 Meaningful Edges

Since level lines capture all details in the image, some of them may be due to noise. Visualizing all level lines at once reveals just an unreadable topographic map. A mechanism for distinguishing important level lines from spurious ones is needed to make sense of the visual information. Good criteria for this are contrast and length. The larger they are, the most information they carry. Desolneux *et al.* [25] have proposed a parameter-free detection of meaningful level lines in a probabilistic sense.

Supposing that n independent measurements of local contrast along a level line can be achieved, considering them as realizations of n random variables of same probability density function ϕ, the probability of their minimum being at least c, the actually observed minimum, is $(\int_c^{+\infty} \phi(x)\,dx)^n$. The result of multiplying this probability by N, the number of level lines, yields the number of false alarms of the line. It can be shown that in an image realization of a

Fig. 8.8 From left to right, intersection of images of Fig. 8.3, and images of Fig. 8.4, and images of Fig. 8.5.

Fig. 8.9 Meaningful level lines of an image. Top: Original image and its level lines (quantized every 10 levels). Bottom: image reconstructed from meaningful level lines and all meaningful level lines (1% of all level lines).

Gaussian noise, the expected number of level lines of NFA less than 1 is 1. In other words in the observed image, only one of the level lines of NFA ≤ 1 is expected to be due to Gaussian noise. Such level lines are called meaningful.

Due to blur, accumulation of meaningful level lines are to be expected along edges. To select only one of them, we select in monotone sections of the tree of *meaningful* level lines the most meaningful one, that is the one with lowest NFA. These are called maximal meaningful.

In absence of a model for ϕ, the empirical histogram of local constrasts may be computed and normalized to yield a probability density function. Local contrast may be defined as the module of gradient, and the length n of a level line is its length in pixels. An example of the capture of the essential geometry by the meaningful level lines is illustrated in Fig. 8.9.

8.3.3 Corner Detection

Partial occlusions of the 3-dimensional scene from the camera position produce T-junctions in the image [17]. Inversely, many T-junctions are a hint of partial occlusion and therefore yield a qualitative information about the 3-D geometry. A T-junction is the phenomenon of two level lines tangent on some extent and then diverging, one of them exhibiting a high curvature at the point of divergence. This is what can be called a corner. Tradional corners, as from Harris and Stephens's detector [41], are just the local maxima of some differerential operator on the image and have no geometry associated. A new definition of corner on a level line by level line basis was introduced by Cao [16].

In this definition, not all local maxima of curvature along a level line are considered corners, but only those that are between two good continuations on the level line. Good continuations are defined as portions of a level line that have low curvature and quantified by an *a contrario* algorithm. Considering a segment of level line of length n pixels we have n measures of curvature spanning an interval of length k in $\mathbb{R}/2\pi\mathbb{Z}$. If the curvatures are independent random variables of uniform law, the probability of this event is $(k/(2\pi))^n$. Multiplying this by the number of possible segments of level lines (i.e., the sum of squares of level line lengths), we get the number of false alarms (NFA) of the segment of level line. Only those of NFA < 1 are meaningful good continuations. Then if two good continuations on a level line are separated by just a few pixels along the level line, a corner is placed in between. In practice, only meaningful level lines are considered for good continuation and corner extraction. An example of corner detection is in Fig. 8.10.

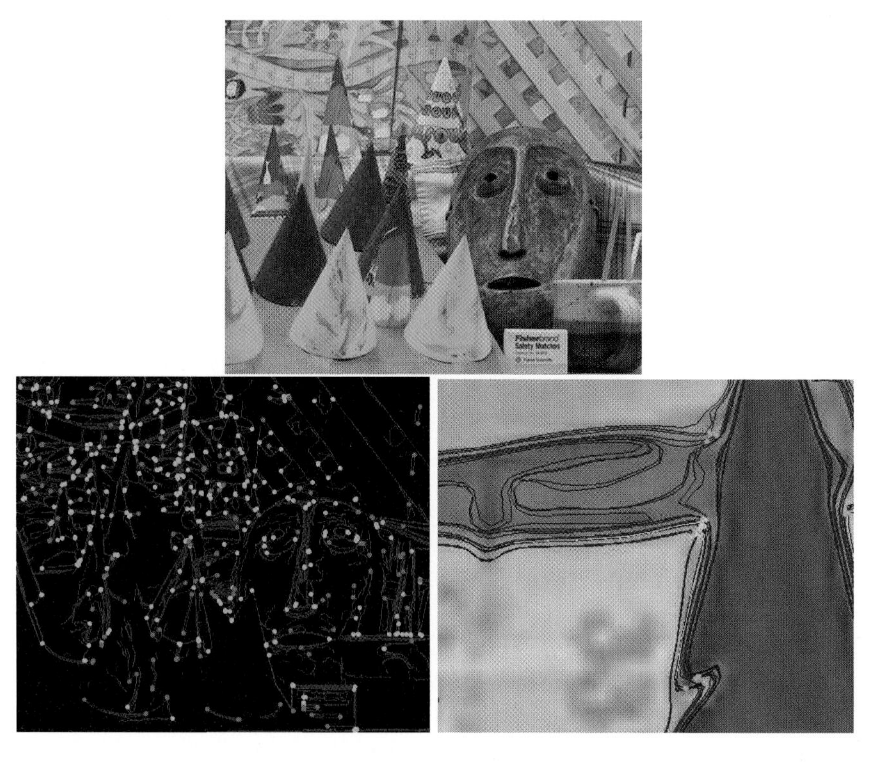

Fig. 8.10 Meaningful corners detection. Top: test image. Bottom left: meaningful corners and good continuations. Bottom right: One particular corner and its two associated good continuations.

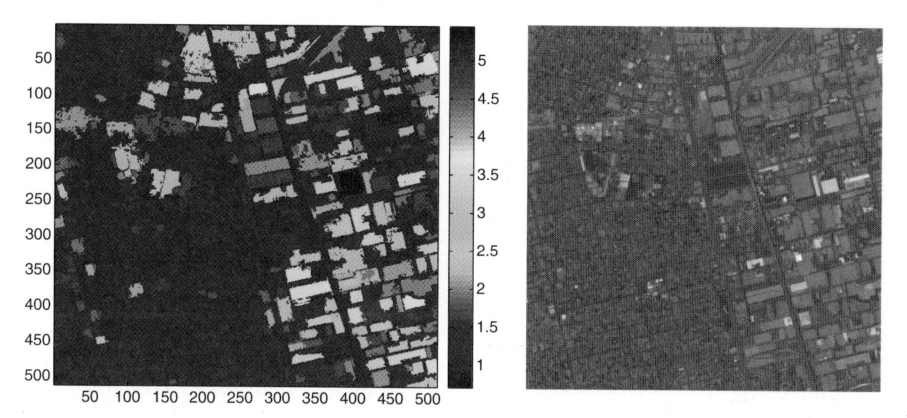

Fig. 8.11 Map of local scales (Left: courtesy of Bin Luo and coauthors [62]) of SPOT5 satellite image of Los Angeles (Right: 5m resolution, copyright CNES).

8.3.4 Scale Adaptive Neighborhoods

A definition for the local scale at pixels of an image based on the tree of shapes was provided by Luo *et al.* [62]. Its advantage compared to linear algorithms and methods based on total variation is its excellent spatial localization. It relies on the most contrasted shape containing a pixel. The definition of contrast is robust to the quantization step and is based on cumulated contrast between similar containing shapes. A parent shape P is said to be similar to its child C if

$$|P| - |C| \leq \lambda Perimeter(C).$$

λ is a parameter depending on level of blur and quantization step, chosen as $\lambda = 1$ by default. This criterion does not impose that C is the only child of P, so groups of similar shapes can cross monotone sections. Successive contrasts in a chain of similar shapes are added to define the contrast of a shape. To each pixel is then associated the most contrasted of its containing shapes, and the local scale of the pixel is the ratio between surface and perimeter of this shape. The map of local scale for a high resolution satellite image is shown in Fig. 8.11.

8.4 Maximally Stable Extremal Regions

The problem of automatic reconstruction of 3D scenes using images taken from arbitrary viewpoints by a series of cameras requires finding reliable correspondences between points or other elements. This problem was addressed in [68] from a point of view much related to the ideas discussed in this monograph. The authors introduced what they called Maximally Stable Extremal Regions and those are the key elements to put into correspondence. Our purpose here is to review the definition of those elements and observe that the computation is immediate using the trees of upper and lower level sets. We also notice that an analogous notion can be defined which is more adapted to the tree of shapes.

Two crucial observations in [68] lead to the choice of Maximally Stable Extremal Regions as robust elements to establish correspondences. In the wide-baseline stereo problem, local image deformations cannot be realistically approximated by euclidian motions and a full affine model is required (as an approximation to the projective transformation between the images). On the other hand, the elements should be robust against illumination changes modeled here as monotonic transformations of image intensities.

With these two observations, the authors of [68] defined the notion of Maximally Stable Extremal Regions. We give it here using the language of the present monograph. Let $u : \overline{\Omega} \to \mathbb{R}$ be a given image (model it as an upper semicontinuous function). We fix a threshold value $\delta > 0$. If u takes values in a set of integers we may take $\delta = 1$. This is the more relevant case in applications since we deal with digital (sampled and quantized) images. For each connected component of an upper level set of u $X_\lambda \in \mathcal{CC}([u \geq \lambda])$ we consider $X_{\lambda-\delta} \in \mathcal{CC}([u \geq \lambda - \delta])$, $X_{\lambda+\delta} \in \mathcal{CC}([u \geq \lambda + \delta])$ such that $X_{\lambda+\delta} \subseteq X_\lambda \subseteq X_{\lambda-\delta}$. We define the function

$$F_\delta^u(\lambda) = \frac{|X_{\lambda-\delta}| - |X_{\lambda+\delta}|}{|X_\lambda|}.$$

We say that X_λ is a Maximally Stable Extremal (upper) Region if X_λ achieves a local minimum of $F_\delta(\lambda)$. The function F_δ is well defined on the maximal branches of the tree of connected components of upper level sets of u, $\mathcal{U}(u)$, where no bifurcation takes place. When X_λ contains a bifurcation, the connected component $X_{\lambda+\delta} \in \mathcal{CC}([u \geq \lambda + \delta])$ is not uniquely defined. We have dismissed those elements.

In a similar way we can define a Maximally Stable Extremal (lower) Region using this time the connected components of the lower level sets of u. We shall refer to both of them as MSER.

In the discussion that follows, we dismiss the problems caused by boundary effects. If A is the affine transformation given by $Ax = Mx + x_0$ where M is a 2×2 matrix and $x_0 \in \mathbb{R}^2$, and we define $u_A(x) = u(Ax)$, then the trees of u and u_A have the same structure. Since the function $F_\delta^u(\lambda) = F_\delta^{u_A}(\lambda)$, we have that the MSER of u and u_A coincide. Thus, they are invariant under affine transformations. Its invariance under contrast changes comes from the fact that we are using connected components of level sets.

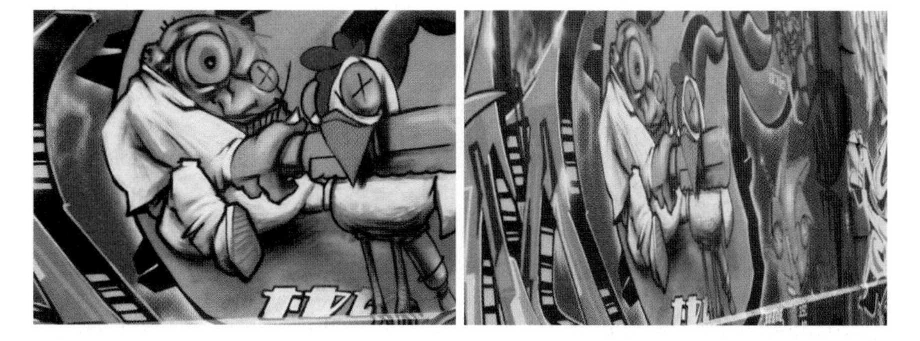

Fig. 8.12 A pair of stereo images taken from http://www.robots.ox.ac.uk/vgg/research/affine/.

To this class of sets we may add a new one, the Maximally Stable Shapes (MSS). They are defined in the same way, this time we take X_0 to be a shape in the tree $\mathcal{S}(u)$ and we take the previous X_{-1} and next X_1 shapes. Those notions take into account that shapes are indexed by their defining gray level λ (in the discrete case, as the saturation of a component of $[u \geq \lambda]$ or of $[u \leq \lambda]$) and if a shape is repeated then its nearby shapes may coincide with it. Then we compute $G^u_\delta(X) = \frac{|X_1|-|X_{-1}|}{|X_0|}$. We say that X is a Maximally Stable Shape if X is a local minimum of the function $G^u_\delta(X)$ defined in the maximal branches of the tree.

We display the MSER (upper and lower) and the MSS corresponding to a pair of stereo images displayed in Fig. 8.12. The top (resp. middle) of Fig. 8.13 displays the upper (resp. lower) MSER computed using the tree of connected components of upper (resp. lower) level sets of u. They are superposed on the original image. Both systems of curves are displayed together in Fig. 8.14. Observe that the upper (resp. lower) MSER surround clear (resp. dark) regions. The bottom of Fig. 8.13 displays (superposed on the original images) the MSS computed using the tree shapes of u. This system of curves is displayed again in Fig. 8.15. They are an alternative to the upper and lower MSER together. We can appreciate that despite the strong projective transformation between both images (which can only be approximated by an affine map) the are many common MSER between the left and right images of Fig. 8.12. MSER, MSS, or the set of meaningful curves provide us with a subset of all level lines of the image which can be used to compute correspondences which in turn can be used for camera calibration or as features to be matched in both images in order to identify them (or find logos).

There is an extensive literature on the subject of feature extraction and image comparison and we do not intend to be exhaustive here. For further reference, we refer to [68, 74] and the references therein. The web page http://www.robots.ox.ac.uk/~vgg/research/affine/ contains many references and a database of images for experimentation.

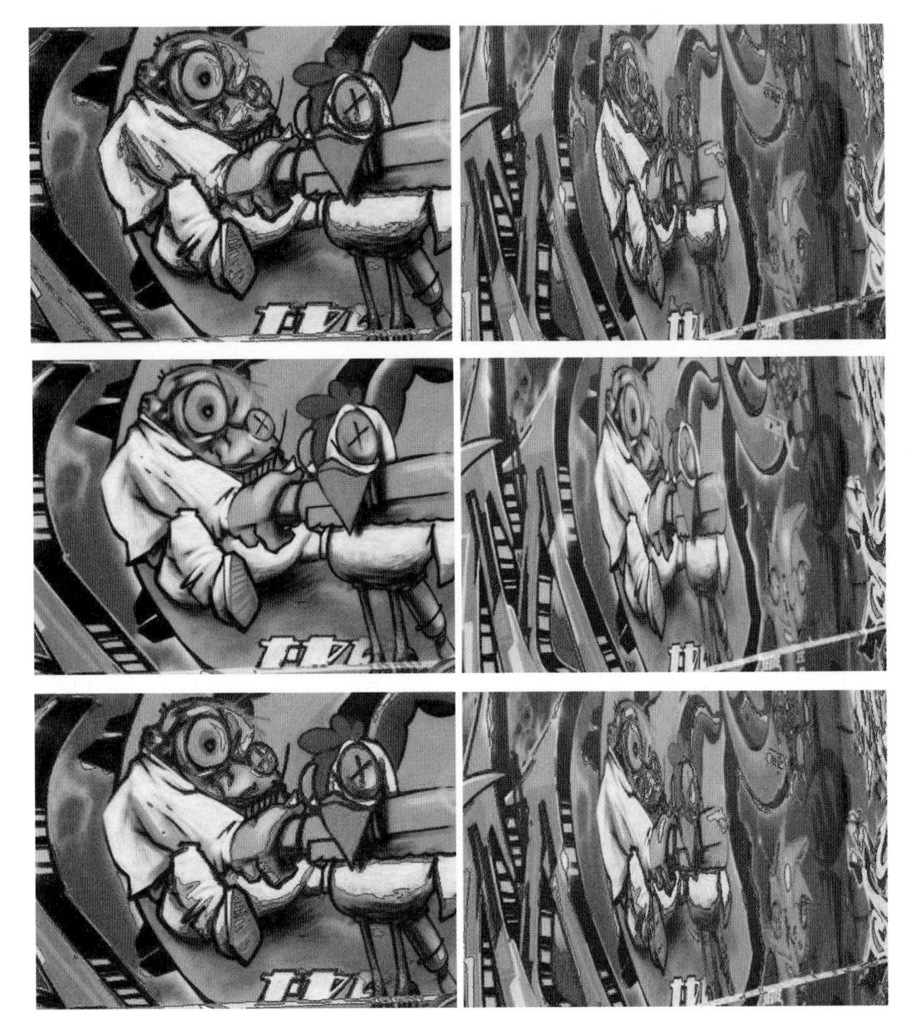

Fig. 8.13 The top images display the MSER boundaries (superposed on the image) corresponding to the two images of Fig. 8.12 computed using the tree $\mathcal{U}(u)$. The middle images display the MSER boundaries computed using the tree $\mathcal{L}(u)$. The bottom images display the Maximally Stable Shape boundaries computed using the tree of shapes $\mathcal{S}(u)$. These images are courtesy of Enric Meinhardt.

Fig. 8.14 We display the MSER boundaries for the two images of Fig. 8.12 corresponding to the upper and lower level set trees. For each image, the boundaries are the union of the curves displayed in the corresponding top and middle images in Fig. 8.13.

Fig. 8.15 We display the Maximally Stable Shape boundaries for the two images of Fig. 8.12 corresponding to the tree of shapes. For each image, the curves coincide with the ones displayed in the corresponding bottom images in Fig. 8.13.

Conclusion

Summary

The tree of shapes of an image, as a unified mix of the component trees, was presented and analyzed in the present notes. Its existence was proven for a semicontinuous image. It relies on the simple model according to which shapes have mostly uniform reflectance properties and are distinguishable from the background. This model has no parameters and is well-posed, though it cannot really account for occlusions and subjective contours notably. The classical area openings of mathematical morphology have their counterpart in the grain filter, which has the advantage of being self-dual on continuous images. If small grains, or small shapes, are filtered out, the tree has a structure essentially finite, and a weak Morse theory was developed under these assumptions: several notions of singularity or criticality were defined and shown to be equivalent. To close the theoretical part, the procedure merging the component trees to build the tree of shapes was detailed, providing an algorithm valid in any dimension. In the most frequent two-dimensional case, a more efficient algorithm, the Fast Level Set Transform, was described. Though not as fast as an optimized component tree algorithm, the FLST is quite usable. A variant based on a bilinear interpolation of the image was also presented, which has the advantage of exhibiting more regular level lines. However, it is slightly slower than the FLST and needs as parameters the levels of quantization to represent the image in a finite structure. Finally, a wide range of concrete applications was exposed and briefly explained.

Extensions

Bilevel Sets

As has been discussed, the tree of shapes can be a better explicit representation of the image than the component trees. However, some important objects are not represented, many of them being bilevel sets, defined by lower *and* upper thresholds. These can be recovered by taking differences of nested shapes, as for example connected components of private pixels of a shape. The resulting sets can be incorporated in the tree. Pushing this idea further, we could also consider differences between a shape and a number of its descendants. As the combinatorial complexity of this can be too expensive, a more structured approach may be needed.

Self-Dual Filters

Whereas some operators, like the median filter, have no natural representation in terms of the tree of shapes, we have seen that the grain filter is a paradigm for *connected* self-dual filters. Keshet [48] brought forth a plurality of other self-dual operators: self-dual erosions and openings. As these operators are only defined in the discrete case, it would be interesting to find a formulation in the continuous case and to study its properties in that framework. In particular, a continuous equivalent to the unfolding transform could be investigated. A good candidate for this would use the minimal total variation along a path originating from the shape and reaching the point at infinity p_∞.

Fast Level Set Transform in Dimension 3

Applying the FLST to 3D images is not very efficient: for each component, we need to compute its Euler characteristic. Whereas it is updated locally in 2D, the locality property is lost in 3D. So the cost of this step in linear in the number of pixels of the component. The merging algorithm of Chap. 5 is better in this regard. It requires the detection of an exterior boundary pixel of components, which is also linear in the number of pixels of the component, but only for upper components of maximal branches in the component tree. A better alternative would be a level surface oriented approach, extending to 3D the algorithm of [109]. Indeed, this does not require the computation of the Euler characteristic. While it also involves following level surfaces, the complexity is only the area of the level surface, not the volume of the component, as a face of a boundary voxel is known right away during the algorithm.

Color Images

An extension of the tree of shapes to color images seems unlikely at first sight. Indeed, the definition relies essentially on level sets, therefore on a total order in the codomain, namely $\mathbb{R}$. As no perceptually relevant order for color codomain $\mathbb{R}^3$ can be imagined, the definition of shapes seems not applicable. Nevertheless, color MSER, or equivalently color component trees, were successfully defined in [33]. This could also be applied to color shapes.

Glossary

$C(D)$	Set of continuous functions on D, 39
G_ε	Grain filter of area ε, 52
$M_\varepsilon, M'_\varepsilon$	Area opening operators of area ε, 40
$NFA(S \leftrightarrow S')$	Number of false alarms of a match, 159
S_x	Smallest limit node containing point x, 21
$[A, B]$	Interval of a tree between nodes A and B, 20
LS_x	The set of level shapes at point x, 24
$\mathrm{cc}(A, x)$	Connected component of A containing point x 11
$\eta_-(x), \eta_+(x)$	end-points of the interval associated to the maximal monotone section containing point x, 80
$\mathcal{L}(u)$	Set of lower components of image u, 18
$\mathcal{M}(S)$	Maximal monotone section containing shape S 24
$\mathcal{P}(X)$	Set of all subsets of X, 18
$\mathcal{S}(u)$	Set of shapes of image u, 18
$\mathcal{T}$	A tree, 18
$\mathcal{U}(u)$	Set of upper components of image u, 18
$\overline{\Omega}$	Domain of image (a set homeomorphic to the closed unit ball of $\mathbb{R}^N$), 11
$\mathrm{Sat}(A, p_\infty)$	Saturation of the set A (w.r.t. point p_∞), 12
$m_{i\,j}, \mu_{i\,j}$	moments and centered moments of a shape, 157
p_∞	Point "at infinity" (a point of $\overline{\Omega}$), 12
$sig(X)$	Signature of set X, 92
$sig(\lambda)$	Signature at level λ, 92
$\mathcal{CC}(A)$	Set of connected components of the point set A 11
$\mathcal{E}$	Set of regional extrema of a function, 91
$\mathcal{USC}(D)$	Set of upper semicontinuous functions on D, 39

References

1. A.V. Aho, J.E. Hopcroft, and J. Ullman. *Data Structures and Algorithms*. Addison-Wesley, 1983.
2. Z. Aktouf, G. Bertrand, and L. Perroton. A three-dimensional holes closing algorithm. *Pattern Recognition Letters*, 23(5):523–531, 2002.
3. L. Alvarez, Y. Gousseau, and J.M. Morel. Scales in natural images and a consequence on their BV norm. In *Proc. of the* 2nd *Workshop on Scale-Space Theories in Computer Vision*, pages 247–258, Corfu, Greece, 1999.
4. L. Alvarez, F. Guichard, P.L. Lions, and J.M. Morel. Axioms and fundamental equations of image processing. *Arch. Rational Mechanics and Anal.*, 16(IX): 200–257, 1993.
5. L. Alvarez, F. Guichard, P.L. Lions, and J.M. Morel. Axioms and fundamental equations of image processing: Multiscale analysis and P.D.E. *Archive for Rational Mechanics and Analysis*, 16(9):200–257, 1993.
6. L. Ambrosio, V. Caselles, S. Masnou, and J.M. Morel. The connected components of sets of finite perimeter and applications to image processing. *European Journal of Mathematics*, 3:39–92, 2001.
7. L. Ambrosio, N. Fusco, and D. Pallara. Functions of bounded variation and free discontinuity problems. *Oxford Mathematical Monographs*, 2000.
8. C. Bajaj, V. Pascucci, and D.R. Schikore. Fast isocontouring for improved interactivity. *In Proc. IEEE Symposium on Volume Visualization, San Francisco, Oct. 7-8*, pages 39–46, 1996.
9. C. Bajaj, V. Pascucci, and D.R. Schikore. The contour spectrum. *Proceedings Visualization '97*, pages 167–173, 1997.
10. C. Ballester and V. Caselles. The m-components of level sets of continuous functions in wbv. *Publicacions Matemàtiques*, 45:477–527, 2001.
11. C. Ballester, V. Caselles, and P. Monasse. The tree of shapes of an image. *ESAIM: Control, Optimisation and Calculus of Variations*, 9:1–18, 2002.
12. C. Ballester, E. Cubero-Castan, M. Gonzalez, and J.M. Morel. Image intersection and applications to satellite imaging. *preprint, C.M.L.A., Ecole Normale Supérieure de Cachan*, 1998.
13. C. Berger, T. Geraud, R. Levillain, N. Widynski, A. Baillard, and E. Bertin. Effective component tree computation with application to pattern recognition in astronomical imaging. In *IEEE Int. Conf. Image Processing, San Antonio, TX*, 2007.
14. P.J. Besl and R.C. Jain. Segmentation through symbolic surface descriptions. *CVPR, May*, 1986.
15. P.J. Burt and E.H. Adelson. A multiresolution spline with application to image mosaics. *ACM Transactions on Graphics*, 2(4):217–236, 1983.
16. F. Cao. Application of the Gestalt principles to the detection of good continuations and corners in image level lines. *Computing and Visualization in Science*, 7(1):3–13, 2004.

17. V. Caselles, B. Coll, and J.M. Morel. A Kanisza program. *Progress in Nonlinear Differential Equations and their Applications*, 25:35–55, 1996.
18. V. Caselles, B. Coll, and J.M. Morel. Topographic maps and local contrast changes in natural images. *Int. J. Comp. Vision*, 33(1):5–27, 1999.
19. V. Caselles, J.L. Lisani, J.M. Morel, and G. Sapiro. Shape preserving local histogram modification. *IEEE Transactions on Image Processing*, 8(2):220–230, 1999.
20. V. Caselles, E. Meinhardt, and P. Monasse. Constructing the tree of shapes of an image by fusion of the trees of connected components of upper and lower level sets. *Positivity, Birkhäuser*, 12(1):55–73, 2008.
21. V. Caselles, G. Sapiro, A. Solé, and C. Ballester. Morse description and morphological encoding of continuous data. *SIAM Journal on Multiscale Modeling and Simulation*, 2(2):179–209, 2004.
22. T.F. Chan, G.H. Golub, and P. Mulet. A nonlinear primal-dual method for total variation based image restoration. *SIAM J. Scientific Computing*, 20(6): 1964–1977, 1999.
23. J.L. Cox, D.B. Karron, and N. Ferdous. Topological zone organization of scalar volume data. *Journal of Mathematical Imaging and Vision*, 18(2):95–117, 2003.
24. E. De Giorgi and L. Ambrosio. Un nuovo tipo di funzionale del calcolo delle variazioni. *Atti Accad. Naz. Lincei*, 8(82):199–210, 1988.
25. A. Desolneux, L. Moisan, and J.M. Morel. Edge detection by Helmholtz principle. *Journal of Mathematical Imaging and Vision*, 14(3):271–284, 2001.
26. A. Desolneux, L. Moisan, and J.M. Morel. *Gestalt theory and image analysis, a probabilistic approach.* Interdisciplinary Applied Mathematics series, Springer Verlag, 2008.
27. F. Dibos and G. Koepfler. Total variation minimization by the fast level sets transform. *Proc. of IEEE Workshop on Variational and Level Sets Methods in Computer Vision*, 2001.
28. F. Dibos, G. Koepfler, and P. Monasse. Image registration. In S. Osher and N. Paragios eds., editors, *Geometric Level Sets Methods in Imaging, Vision and Graphics*, 2002.
29. F. Dibos, G. Koepfler, and P. Monasse. Total variation minimization: application to gray-scale, color images and optical flow regularization. In S. Osher and N. Paragios eds., editors, *Geometric Level Sets Methods in Imaging, Vision and Graphics*, 2002.
30. S. Durand, F. Malgouyres, and B. Rougé. Image deblurring, spectrum interpolation and application to satellite imaging. *Mathematical Modelling and Numerical Analysis*, 1999.
31. L.C. Evans and R.F. Gariepy. Measure theory and fine properties of functions. *Studies in Advanced Math., CRC Press*, 1992.
32. M.A. Fischler and C. Bolles. Random Sample Consensus: A paradigm for model fitting with applications to image analysis and automated cartography. *Communications of the ACM*, 24:381–395, 1981.
33. Per-Erik Forssén. Maximally stable colour regions for recognition and matching. In *IEEE Conference on Computer Vision and Pattern Recognition*, 2007.
34. W.R. Franklin and A. Said. Lossy compression of elevation data. *7th Int. Symposium on Spatial Data Handling*, 1996.
35. J. Froment. A compact and multiscale image model based on level sets. In *Proc. of the* 2nd *Workshop on Scale-Space Theories in Computer Vision*, pages 152–163, Corfu, Greece, 1999.
36. M. Golubitsky and V. Guillemin. *Stable Mappings and Their Singularities.* Springer Verlag, 1973.
37. Y. Gousseau and J.M. Morel. Texture synthesis through level sets. *Preprint CMLA*, 2000.

38. J. Goutsias and H.J.A.M. Heijmans. Fundamenta morphologicae mathematicae. *Fundamenta Informaticae*, 41(1-2):1–31, 2000.
39. F. Guichard and J.M. Morel. Partial differential equations and image iterative filtering. *Forthcoming book*, 2003.
40. R. Haralick, L. Winston, and T. Laffey. The topographic primal sketch. *Int. J. Rob. Research*, 2, 1983.
41. C. Harris and M. Stephens. A combined corner and edge detector. In *Proceedings of the 4th Alvey Vision Conference*, pages 147–151, 1988.
42. H. Heijmans and J. Serra. Convergence, continuity and iteration in mathematical morphology. *Journal of Visual Communication and Image Representation*, 3:84–102, 1992.
43. H.J.A.M. Heijmans. *Morphological Image Operators*. Academic Press, Boston, Mass., 1994.
44. H.J.A.M. Heijmans. Self-dual morphological operators and filters. *J. Math. Imaging and Vision*, 6(1):15–36, 1996.
45. H.J.A.M. Heijmans. Connected morphological operators for binary images. In *IEEE Int. Conference on Image Processing, ICIP'97*, volume 2, pages 211–214, Santa barbara (CA), USA, October 1997.
46. K. Hormann, S. Spinello, and P. Schröder. c^1-continuous terrain reconstruction from sparse contours. In T. Ertl, B. Girod, G. Greiner, H. Niemann, H.P. Seidel, E. Steinbach, and R. Westermann, editors, *Proc. 8th Int. Worksh. Vision, Modeling, and Visualization*, pages 289–297. IOS Press, 2003.
47. E.G. Johnston and A. Rosenfeld. Digital detection of pits, peaks, ridges and ravines. *IEEE Trans. Systems Man Cybernetics*, 472(July), 1975.
48. R. Keshet. Adjacency lattices and shape-tree semilattices. *Image and Vision Computing*, 25(4):436–446, 2007.
49. T.Y. Kong and A. Rosenfeld. If we use 4- or 8-connectedness for both the objects and the background, the Euler characteristic is not locally computable. *Pattern Recognition Letter*, 11:231–232, 1990.
50. A.S. Kronrod. On functions of two variables. *Uspehi Mathematical Sciences (NS)*, 35(5):24–134, 1950.
51. C. Kuratowski. *Topologie I, II*. Editions J. Gabay, Paris, 1992.
52. I.S. Kweon and T. Kanade. Extracting topograpic terrain features from elevation maps. *CVGIP: Image Understanding*, 59(2):171–182, 1994.
53. C. Lantuéjoul and S. Beucher. On the use of geodesic metric in image analysis. *J. Microscopy*, 121:39–49, 1981.
54. C. Lantuéjoul and F. Maisonneuve. Geodesic methods in image analysis. *Pattern Recognition*, 17:117–187, 1984.
55. C.N. Lee, T. Poston, and A. Rosenfeld. Holes and genus of 2d and 3d digital images. *Graphical Models and Image Processing*, 55(1):20–yy, January 1993.
56. C.N. Lee and A. Rosenfeld. Computing the Euler number of a 3d image. In *Proceedings of International Conference of Computer Vision*, pages 567–571, 1987.
57. J.L. Lisani. Comparaison automatique d'images par leurs formes. *Ph.D Thesis, Université de Paris-Dauphine, July*, 2001.
58. J.L. Lisani, L. Moisan, P. Monasse, and J.M. Morel. Affine invariant mathematical morphology applied to a generic shape recognition algorithm. *Comp. Imaging and Vision*, 18, 2000.
59. D.G. Lowe. Object recognition from local scale-invariant features. In *Proc. of the International Conference on Computer Vision*, pages 1150–1157, 1999.
60. R. Lumia. A new three-dimensional connected components algorithm. *Computer Vision, Graphics and Image Processing*, 23(2):207–217, August 1983.
61. R. Lumia, L.G. Shapiro, and O.A. Zuniga. A new connected components algorithm for virtual memory computers. *Computer Vision, Graphics and Image Processing*, 22(2):287–300, May 1983.

62. B. Luo, J.F. Aujol, and Y. Gousseau. Local scale measure from the topographic map and application to remote sensing images. Preprint CMLA, 2008.
63. S. Mallat. *A Wavelet Tour of Signal Processing.* Academic Press, New York, 1998.
64. P. Maragos and R.W. Schafer. Morphological filters. part I: Their set-theoretic analysis and relations to linear shift-invariant filters. *IEEE Transactions on Acoustics, Speech and Signal Processing*, 35:1153–1169, 1987.
65. P. Maragos and R.W. Schafer. Morphological filters. part II: Their relations to median, order-statistic, and stack filters. *IEEE Transactions on Acoustics, Speech and Signal Processing*, 35:1170–1184, 1987.
66. D. Marr. Vision. *Freeman and Co.*, 1981.
67. S. Masnou. Image restoration involving connectedness. In *Proceedings of the* 6th *International Workshop on Digital Image Processing and Computer Graphics*, volume 3346, Vienna, Austria, 1998. SPIE.
68. J. Matas, O. Chum, M. Urban, and T. Pajdla. Robust wide baseline stereo from maximally stable extremal regions. In *Proc. of British Machine Vision Conference*, pages 384–393, 2002.
69. G. Matheron. *Random Sets and Integral Geometry.* John Wiley, N.Y., 1975.
70. F. Meyer. *Mathematical Morphology and Its Application to Signal and Image Processing*, chapter From Connected Operators to Levelings. Kluwer Academic Publishers, H.J.A.M. Heijmans and J. Roerdink edition, 1998.
71. F. Meyer. *Mathematical Morphology and Its Application to Signal and Image Processing*, chapter The Levelings. Kluwer Academic Publishers, H.J.A.M. Heijmans and J. Roerdink edition, 1998.
72. F. Meyer and S. Beucher. Morphological segmentation. *Journal of Visual Communication and Image Representation*, 1(1):21–46, September 1990.
73. F. Meyer and P. Maragos. Morphological scale-space representation with levelings. In *Proc. of the* 2nd *Workshop on Scale-Space Theories in Computer Vision*, pages 187–198, Corfu, Greece, 1999.
74. K. Mikolajczyk, T. Tuytelaars, C. Schmid, A. Zisserman, J. Matas, F. Schaffalitzky, T. Kadir, and L. Van Gool. A comparison of affine region detectors. *International Journal of Computer Vision*, 65(1/2):43–72, 2005.
75. J. Milnor. Morse theory. *Annals of Math. Studies 51, Princeton University Press*, 1963.
76. L. Moisan. Affine plane curve evolution: A fully consistent scheme. *IEEE Transactions on Image Processing*, 7(3):411–420, March 1998.
77. P. Monasse. Contrast invariant image registration. In *Proc. of International Conference on Acoustics, Speech and Signal Processing, 6*, pages 3221–3224, 1999.
78. P. Monasse. *Représentation morphologique d'images numériques et application au recalage.* PhD thesis, Université de Paris-Dauphine, 2000.
79. P. Monasse and F. Guichard. Fast computation of a contrast-invariant image representation. *IEEE Transactions on Image Processing*, 9(5):860–872, 2000.
80. P. Monasse and F. Guichard. Scale-space from a level lines tree. *Journal of Visual Communication and Image Representation*, 11:224–236, 2000.
81. J.M. Morel and S. Solimini. Variational methods in image processing. *Birkhäuser*, 1994.
82. D. Mumford and J. Shah. Optimal approximations by piecewise smooth functions and associated variational problems. *Communications on Pure and Applied Mathematics*, 17:577–585, 1989.
83. L. Najman and M. Couprie. Building the component tree in quasi-linear time. *IEEE Transactions on Image Processing*, 15(11):3531–3539, 2006.
84. M.H.A. Newman. Elements of the topology of plane sets of points. *Dover Publications, New York*, 1992.

85. D. Nistér and H. Stewénius. Linear time maximally stable extremal regions. In *Proc. of ECCV*, LNCS 5503, pages 183–196, 2008.
86. Y. Ohta and T. Kanade. Stereo by intra- and inter-scanline search using dynamic programming. *IEEE Transactions on Pattern Analysis and Machine Intelligence*, 7:139–154, 1985.
87. G. Reeb. Sur les poits singuliers d'une forme de pfaff completement integrable ou d'une fonction numérique. *Comptes Rendus Acad. Sciences Paris*, 222:847–849, 1946.
88. A. Rosenfeld and A.C. Kak. *Digital Picture Processing.* Academic Press, 1982.
89. A. Rosenfeld and J.L. Pfaltz. Sequential operations in digital picture processing. *Journal of Applied and Computational Mathematics*, 13(4):471–494, October 1966.
90. J. Roubal and T.K. Poiker. Automated contour labelling and the contour tree. *In Auto-Carto, March*, 1985.
91. L. Rudin, S. Osher, and E. Fatemi. Nonlinear total variation based noise removal algorithms. *Physica D*, 60:259–268, 1992.
92. P. Salembier. Morphological multiscale segmentation for image coding. *Signal Processing, Special Issue on Nonlinear Signal Processing*, 38(3):359–386, 1994.
93. P. Salembier. Region-based filtering of images and video sequences: A morphological viewpoint. *IEEE Trans. on Circuits and Systems for Video Technology*, 9(8):1147–1167, 1999.
94. P. Salembier, P. Brigger, J.R. Casas, and M. Pardàs. Morphological operators for image and video compression. *IEEE Trans. on Image Processing*, 5(6):881–897, June 1996.
95. P. Salembier and L. Garrido. Binary partition tree as an efficient representation for image processing, segmentation, and information retrieval. *IEEE Transactions on Image Processing*, 9(4):561–576, 2000.
96. P. Salembier, F. Meyer, P. Brigger, and L. Bouchard. Morphological operators for very low bit rate video coding. In *Proceedings of International Conference of Image Processing*, page 19P1, 1996.
97. P. Salembier, A. Oliveras, and L. Garrido. Antiextensive connected operators for image and sequence processing. *IEEE Transactions on Image Processing*, 7(4):555–570, April 1998.
98. P. Salembier and J. Serra. Flat zones filtering, connected operators and filters by reconstruction. *IEEE Trans. Image Processing*, 4:1153–1160, 1995.
99. G. Sapiro and A. Tannenbaum. Affine invariant scale-space. *International Journal of Computer Vision*, 11(1):25–44, 1993.
100. G. Sapiro and A. Tannenbaum. On affine plane curve evolution. *Journal of Functional Analysis*, 119(1):79–120, 1994.
101. R. Sedgewick. *Algorithms in C++: Fundamentals, Data Structures, Sorting, Searching.* Addison-Wesley, 3 edition, 1999.
102. J. Serra. *Image Analysis and Mathematical Morphology.* Academic Press, New York, 1982.
103. J. Serra. Introduction to mathematical morphology. *Computer Vision, Graphics and Image Processing*, 35(3):283–305, September 1986.
104. J. Serra. *Image Analysis and Mathematical Morphology 2: Theoretical Advances.* Academic Press, New York, 1988.
105. J. Serra. Connections for sets and functions. *Preprint, École des Mines de Paris, Centre de Morphologie Mathématique, Fontainebleau, France*, May 1999.
106. J. Serra and P. Salembier. Connected operators and pyramids. In *Proceedings of SPIE Conference on Image Algebra and Mathematical Morphology*, volume 2030, pages 65–76, San Diego, California, 1993.
107. Y. Shinagawa, T.L. Kunii, and Y.L. Kergosien. Surface coding based on morse theory. *IEEE Computer Graphics and Appl.*, 11(5):66–78, 1991.

108. A. Solé, V. Caselles, G. Sapiro, and F. Arándiga. Morse description and geometric encoding of digital elevation maps. *IEEE Transactions on Image Processing*, 13(9):1245–1262, 2004.
109. Y. Song. A topdown algorithm for computation of level line trees. *IEEE Transactions on Image Processing*, 16(8):2107–2116, 2007.
110. S. Takahashi, T. Ikeda, Y. Shinagawa, T. Kunii, and M. Ueda. Algorithms for extracting correct critical points and constructing topological graphs from discrete geographic elevation data. *Eurographics 95*, 14(3):181–192, 1995.
111. R.E. Tarjan. Efficiency of a good but not linear set union algorithm. *Journal of the ACM*, 22:215–225, 1975.
112. C. Vachier. Valuation of image extrema using alterning filters by reconstruction. In *Proc. SPIE, Image Algebra and Morphological Processing*, 1995.
113. M. van Kreveld, R. va Oostrum, V. Bajaj, V. Pascucci, and D. Schikore. Contour trees and small seed sets for isosurface traversal. *In 13th ACM Symposium on Computational Geometry*, pages 212–220, 1997.
114. L. Vincent. Grayscale area openings and closings, their efficient implementation and applications. In J. Serra and Ph. Salembrier, editors, *Proceedings of the* 1st *Workshop on Mathematical Morphology and its Applications to Signal Processing*, pages 22–27, Barcelona, Spain, 1993.
115. L. Vincent. Morphological area openings and closings for grey-scale images. In *Proceedings of the Workshop Shape in Picture: Mathematical Description of Shape in Gray-Level Images*, pages 197–208, Driebergen, The Netherlands, 1994. Springer, Berlin.
116. L. Vincent and P. Soille. Watersheds in digital spaces: An efficient algorithm based on immersion simulations. *IEEE Transactions on Pattern Analysis and Machine Intelligence*, 13(6):583–598, June 1991.
117. C. R. Vogel and M. E. Oman. Iterative methods for total variation denoising. *SIAM J. Sci. Computing*, 17:227–238, 1996.
118. M. Wertheimer. Untersuchungen zur Lehre der Gestalt, II. *Psychologische Forschung*, 4:301–350, 1923.
119. L.P. Yaroslavski and M. Eden. *Fundamentals of digital optics*. Birkhäuser, Boston, 1996.

Index

Lecture Notes in Mathematics

For information about earlier volumes
please contact your bookseller or Springer
LNM Online archive: springerlink.com

Vol. 1801: J. Azéma, M. Émery, M. Ledoux, M. Yor (Eds.), Séminaire de Probabilités XXXVI (2003)
Vol. 1802: V. Capasso, E. Merzbach, B. G. Ivanoff, M. Dozzi, R. Dalang, T. Mountford, Topics in Spatial Stochastic Processes. Martina Franca, Italy 2001. Editor: E. Merzbach (2003)
Vol. 1803: G. Dolzmann, Variational Methods for Crystalline Microstructure – Analysis and Computation (2003)
Vol. 1804: I. Cherednik, Ya. Markov, R. Howe, G. Lusztig, Iwahori-Hecke Algebras and their Representation Theory. Martina Franca, Italy 1999. Editors: V. Baldoni, D. Barbasch (2003)
Vol. 1805: F. Cao, Geometric Curve Evolution and Image Processing (2003)
Vol. 1806: H. Broer, I. Hoveijn. G. Lunther, G. Vegter, Bifurcations in Hamiltonian Systems. Computing Singularities by Gröbner Bases (2003)
Vol. 1807: V. D. Milman, G. Schechtman (Eds.), Geometric Aspects of Functional Analysis. Israel Seminar 2000-2002 (2003)
Vol. 1808: W. Schindler, Measures with Symmetry Properties (2003)
Vol. 1809: O. Steinbach, Stability Estimates for Hybrid Coupled Domain Decomposition Methods (2003)
Vol. 1810: J. Wengenroth, Derived Functors in Functional Analysis (2003)
Vol. 1811: J. Stevens, Deformations of Singularities (2003)
Vol. 1812: L. Ambrosio, K. Deckelnick, G. Dziuk, M. Mimura, V. A. Solonnikov, H. M. Soner, Mathematical Aspects of Evolving Interfaces. Madeira, Funchal, Portugal 2000. Editors: P. Colli, J. F. Rodrigues (2003)
Vol. 1813: L. Ambrosio, L. A. Caffarelli, Y. Brenier, G. Buttazzo, C. Villani, Optimal Transportation and its Applications. Martina Franca, Italy 2001. Editors: L. A. Caffarelli, S. Salsa (2003)
Vol. 1814: P. Bank, F. Baudoin, H. Föllmer, L.C.G. Rogers, M. Soner, N. Touzi, Paris-Princeton Lectures on Mathematical Finance 2002 (2003)
Vol. 1815: A. M. Vershik (Ed.), Asymptotic Combinatorics with Applications to Mathematical Physics. St. Petersburg, Russia 2001 (2003)
Vol. 1816: S. Albeverio, W. Schachermayer, M. Talagrand, Lectures on Probability Theory and Statistics. Ecole d'Eté de Probabilités de Saint-Flour XXX-2000. Editor: P. Bernard (2003)
Vol. 1817: E. Koelink, W. Van Assche (Eds.), Orthogonal Polynomials and Special Functions. Leuven 2002 (2003)
Vol. 1818: M. Bildhauer, Convex Variational Problems with Linear, nearly Linear and/or Anisotropic Growth Conditions (2003)
Vol. 1819: D. Masser, Yu. V. Nesterenko, H. P. Schlickewei, W. M. Schmidt, M. Waldschmidt, Diophantine Approximation. Cetraro, Italy 2000. Editors: F. Amoroso, U. Zannier (2003)
Vol. 1820: F. Hiai, H. Kosaki, Means of Hilbert Space Operators (2003)
Vol. 1821: S. Teufel, Adiabatic Perturbation Theory in Quantum Dynamics (2003)
Vol. 1822: S.-N. Chow, R. Conti, R. Johnson, J. Mallet-Paret, R. Nussbaum, Dynamical Systems. Cetraro, Italy 2000. Editors: J. W. Macki, P. Zecca (2003)
Vol. 1823: A. M. Anile, W. Allegretto, C. Ringhofer, Mathematical Problems in Semiconductor Physics. Cetraro, Italy 1998. Editor: A. M. Anile (2003)
Vol. 1824: J. A. Navarro González, J. B. Sancho de Salas, $\mathscr{C}^\infty$ – Differentiable Spaces (2003)
Vol. 1825: J. H. Bramble, A. Cohen, W. Dahmen, Multiscale Problems and Methods in Numerical Simulations, Martina Franca, Italy 2001. Editor: C. Canuto (2003)
Vol. 1826: K. Dohmen, Improved Bonferroni Inequalities via Abstract Tubes. Inequalities and Identities of Inclusion-Exclusion Type. VIII, 113 p, 2003.
Vol. 1827: K. M. Pilgrim, Combinations of Complex Dynamical Systems. IX, 118 p, 2003.
Vol. 1828: D. J. Green, Gröbner Bases and the Computation of Group Cohomology. XII, 138 p, 2003.
Vol. 1829: E. Altman, B. Gaujal, A. Hordijk, Discrete-Event Control of Stochastic Networks: Multimodularity and Regularity. XIV, 313 p, 2003.
Vol. 1830: M. I. Gil', Operator Functions and Localization of Spectra. XIV, 256 p, 2003.
Vol. 1831: A. Connes, J. Cuntz, E. Guentner, N. Higson, J. E. Kaminker, Noncommutative Geometry, Martina Franca, Italy 2002. Editors: S. Doplicher, L. Longo (2004)
Vol. 1832: J. Azéma, M. Émery, M. Ledoux, M. Yor (Eds.), Séminaire de Probabilités XXXVII (2003)
Vol. 1833: D.-Q. Jiang, M. Qian, M.-P. Qian, Mathematical Theory of Nonequilibrium Steady States. On the Frontier of Probability and Dynamical Systems. IX, 280 p, 2004.
Vol. 1834: Yo. Yomdin, G. Comte, Tame Geometry with Application in Smooth Analysis. VIII, 186 p, 2004.
Vol. 1835: O.T. Izhboldin, B. Kahn, N.A. Karpenko, A. Vishik, Geometric Methods in the Algebraic Theory of Quadratic Forms. Summer School, Lens, 2000. Editor: J.-P. Tignol (2004)
Vol. 1836: C. Năstăsescu, F. Van Oystaeyen, Methods of Graded Rings. XIII, 304 p, 2004.
Vol. 1837: S. Tavaré, O. Zeitouni, Lectures on Probability Theory and Statistics. Ecole d'Eté de Probabilités de Saint-Flour XXXI-2001. Editor: J. Picard (2004)
Vol. 1838: A.J. Ganesh, N.W. O'Connell, D.J. Wischik, Big Queues. XII, 254 p, 2004.
Vol. 1839: R. Gohm, Noncommutative Stationary Processes. VIII, 170 p, 2004.

Vol. 1840: B. Tsirelson, W. Werner, Lectures on Probability Theory and Statistics. Ecole d'Eté de Probabilités de Saint-Flour XXXII-2002. Editor: J. Picard (2004)
Vol. 1841: W. Reichel, Uniqueness Theorems for Variational Problems by the Method of Transformation Groups (2004)
Vol. 1842: T. Johnsen, A. L. Knutsen, K_3 Projective Models in Scrolls (2004)
Vol. 1843: B. Jefferies, Spectral Properties of Noncommuting Operators (2004)
Vol. 1844: K.F. Siburg, The Principle of Least Action in Geometry and Dynamics (2004)
Vol. 1845: Min Ho Lee, Mixed Automorphic Forms, Torus Bundles, and Jacobi Forms (2004)
Vol. 1846: H. Ammari, H. Kang, Reconstruction of Small Inhomogeneities from Boundary Measurements (2004)
Vol. 1847: T.R. Bielecki, T. Björk, M. Jeanblanc, M. Rutkowski, J.A. Scheinkman, W. Xiong, Paris-Princeton Lectures on Mathematical Finance 2003 (2004)
Vol. 1848: M. Abate, J. E. Fornaess, X. Huang, J. P. Rosay, A. Tumanov, Real Methods in Complex and CR Geometry, Martina Franca, Italy 2002. Editors: D. Zaitsev, G. Zampieri (2004)
Vol. 1849: Martin L. Brown, Heegner Modules and Elliptic Curves (2004)
Vol. 1850: V. D. Milman, G. Schechtman (Eds.), Geometric Aspects of Functional Analysis. Israel Seminar 2002-2003 (2004)
Vol. 1851: O. Catoni, Statistical Learning Theory and Stochastic Optimization (2004)
Vol. 1852: A.S. Kechris, B.D. Miller, Topics in Orbit Equivalence (2004)
Vol. 1853: Ch. Favre, M. Jonsson, The Valuative Tree (2004)
Vol. 1854: O. Saeki, Topology of Singular Fibers of Differential Maps (2004)
Vol. 1855: G. Da Prato, P.C. Kunstmann, I. Lasiecka, A. Lunardi, R. Schnaubelt, L. Weis, Functional Analytic Methods for Evolution Equations. Editors: M. Iannelli, R. Nagel, S. Piazzera (2004)
Vol. 1856: K. Back, T.R. Bielecki, C. Hipp, S. Peng, W. Schachermayer, Stochastic Methods in Finance, Bressanone/Brixen, Italy, 2003. Editors: M. Fritelli, W. Runggaldier (2004)
Vol. 1857: M. Émery, M. Ledoux, M. Yor (Eds.), Séminaire de Probabilités XXXVIII (2005)
Vol. 1858: A.S. Cherny, H.-J. Engelbert, Singular Stochastic Differential Equations (2005)
Vol. 1859: E. Letellier, Fourier Transforms of Invariant Functions on Finite Reductive Lie Algebras (2005)
Vol. 1860: A. Borisyuk, G.B. Ermentrout, A. Friedman, D. Terman, Tutorials in Mathematical Biosciences I. Mathematical Neurosciences (2005)
Vol. 1861: G. Benettin, J. Henrard, S. Kuksin, Hamiltonian Dynamics – Theory and Applications, Cetraro, Italy, 1999. Editor: A. Giorgilli (2005)
Vol. 1862: B. Helffer, F. Nier, Hypoelliptic Estimates and Spectral Theory for Fokker-Planck Operators and Witten Laplacians (2005)
Vol. 1863: H. Führ, Abstract Harmonic Analysis of Continuous Wavelet Transforms (2005)
Vol. 1864: K. Efstathiou, Metamorphoses of Hamiltonian Systems with Symmetries (2005)
Vol. 1865: D. Applebaum, B.V. R. Bhat, J. Kustermans, J. M. Lindsay, Quantum Independent Increment Processes I. From Classical Probability to Quantum Stochastic Calculus. Editors: M. Schürmann, U. Franz (2005)
Vol. 1866: O.E. Barndorff-Nielsen, U. Franz, R. Gohm, B. Kümmerer, S. Thorbjønsen, Quantum Independent Increment Processes II. Structure of Quantum Lévy Processes, Classical Probability, and Physics. Editors: M. Schürmann, U. Franz, (2005)
Vol. 1867: J. Sneyd (Ed.), Tutorials in Mathematical Biosciences II. Mathematical Modeling of Calcium Dynamics and Signal Transduction. (2005)
Vol. 1868: J. Jorgenson, S. Lang, $Pos_n(R)$ and Eisenstein Series. (2005)
Vol. 1869: A. Dembo, T. Funaki, Lectures on Probability Theory and Statistics. Ecole d'Eté de Probabilités de Saint-Flour XXXIII-2003. Editor: J. Picard (2005)
Vol. 1870: V.I. Gurariy, W. Lusky, Geometry of Müntz Spaces and Related Questions. (2005)
Vol. 1871: P. Constantin, G. Gallavotti, A.V. Kazhikhov, Y. Meyer, S. Ukai, Mathematical Foundation of Turbulent Viscous Flows, Martina Franca, Italy, 2003. Editors: M. Cannone, T. Miyakawa (2006)
Vol. 1872: A. Friedman (Ed.), Tutorials in Mathematical Biosciences III. Cell Cycle, Proliferation, and Cancer (2006)
Vol. 1873: R. Mansuy, M. Yor, Random Times and Enlargements of Filtrations in a Brownian Setting (2006)
Vol. 1874: M. Yor, M. Émery (Eds.), In Memoriam Paul-André Meyer - Séminaire de Probabilités XXXIX (2006)
Vol. 1875: J. Pitman, Combinatorial Stochastic Processes. Ecole d'Eté de Probabilités de Saint-Flour XXXII-2002. Editor: J. Picard (2006)
Vol. 1876: H. Herrlich, Axiom of Choice (2006)
Vol. 1877: J. Steuding, Value Distributions of L-Functions (2007)
Vol. 1878: R. Cerf, The Wulff Crystal in Ising and Percolation Models, Ecole d'Eté de Probabilités de Saint-Flour XXXIV-2004. Editor: Jean Picard (2006)
Vol. 1879: G. Slade, The Lace Expansion and its Applications, Ecole d'Eté de Probabilités de Saint-Flour XXXIV-2004. Editor: Jean Picard (2006)
Vol. 1880: S. Attal, A. Joye, C.-A. Pillet, Open Quantum Systems I, The Hamiltonian Approach (2006)
Vol. 1881: S. Attal, A. Joye, C.-A. Pillet, Open Quantum Systems II, The Markovian Approach (2006)
Vol. 1882: S. Attal, A. Joye, C.-A. Pillet, Open Quantum Systems III, Recent Developments (2006)
Vol. 1883: W. Van Assche, F. Marcellàn (Eds.), Orthogonal Polynomials and Special Functions, Computation and Application (2006)
Vol. 1884: N. Hayashi, E.I. Kaikina, P.I. Naumkin, I.A. Shishmarev, Asymptotics for Dissipative Nonlinear Equations (2006)
Vol. 1885: A. Telcs, The Art of Random Walks (2006)
Vol. 1886: S. Takamura, Splitting Deformations of Degenerations of Complex Curves (2006)
Vol. 1887: K. Habermann, L. Habermann, Introduction to Symplectic Dirac Operators (2006)
Vol. 1888: J. van der Hoeven, Transseries and Real Differential Algebra (2006)
Vol. 1889: G. Osipenko, Dynamical Systems, Graphs, and Algorithms (2006)
Vol. 1890: M. Bunge, J. Funk, Singular Coverings of Toposes (2006)
Vol. 1891: J.B. Friedlander, D.R. Heath-Brown, H. Iwaniec, J. Kaczorowski, Analytic Number Theory, Cetraro, Italy, 2002. Editors: A. Perelli, C. Viola (2006)
Vol. 1892: A. Baddeley, I. Bárány, R. Schneider, W. Weil, Stochastic Geometry, Martina Franca, Italy, 2004. Editor: W. Weil (2007)

Vol. 1893: H. Hanßmann, Local and Semi-Local Bifurcations in Hamiltonian Dynamical Systems, Results and Examples (2007)
Vol. 1894: C.W. Groetsch, Stable Approximate Evaluation of Unbounded Operators (2007)
Vol. 1895: L. Molnár, Selected Preserver Problems on Algebraic Structures of Linear Operators and on Function Spaces (2007)
Vol. 1896: P. Massart, Concentration Inequalities and Model Selection, Ecole d'Été de Probabilités de Saint-Flour XXXIII-2003. Editor: J. Picard (2007)
Vol. 1897: R. Doney, Fluctuation Theory for Lévy Processes, Ecole d'Été de Probabilités de Saint-Flour XXXV-2005. Editor: J. Picard (2007)
Vol. 1898: H.R. Beyer, Beyond Partial Differential Equations, On linear and Quasi-Linear Abstract Hyperbolic Evolution Equations (2007)
Vol. 1899: Séminaire de Probabilités XL. Editors: C. Donati-Martin, M. Émery, A. Rouault, C. Stricker (2007)
Vol. 1900: E. Bolthausen, A. Bovier (Eds.), Spin Glasses (2007)
Vol. 1901: O. Wittenberg, Intersections de deux quadriques et pinceaux de courbes de genre 1, Intersections of Two Quadrics and Pencils of Curves of Genus 1 (2007)
Vol. 1902: A. Isaev, Lectures on the Automorphism Groups of Kobayashi-Hyperbolic Manifolds (2007)
Vol. 1903: G. Kresin, V. Maz'ya, Sharp Real-Part Theorems (2007)
Vol. 1904: P. Giesl, Construction of Global Lyapunov Functions Using Radial Basis Functions (2007)
Vol. 1905: C. Prévôt, M. Röckner, A Concise Course on Stochastic Partial Differential Equations (2007)
Vol. 1906: T. Schuster, The Method of Approximate Inverse: Theory and Applications (2007)
Vol. 1907: M. Rasmussen, Attractivity and Bifurcation for Nonautonomous Dynamical Systems (2007)
Vol. 1908: T.J. Lyons, M. Caruana, T. Lévy, Differential Equations Driven by Rough Paths, Ecole d'Été de Probabilités de Saint-Flour XXXIV-2004 (2007)
Vol. 1909: H. Akiyoshi, M. Sakuma, M. Wada, Y. Yamashita, Punctured Torus Groups and 2-Bridge Knot Groups (I) (2007)
Vol. 1910: V.D. Milman, G. Schechtman (Eds.), Geometric Aspects of Functional Analysis. Israel Seminar 2004-2005 (2007)
Vol. 1911: A. Bressan, D. Serre, M. Williams, K. Zumbrun, Hyperbolic Systems of Balance Laws. Cetraro, Italy 2003. Editor: P. Marcati (2007)
Vol. 1912: V. Berinde, Iterative Approximation of Fixed Points (2007)
Vol. 1913: J.E. Marsden, G. Misiołek, J.-P. Ortega, M. Perlmutter, T.S. Ratiu, Hamiltonian Reduction by Stages (2007)
Vol. 1914: G. Kutyniok, Affine Density in Wavelet Analysis (2007)
Vol. 1915: T. Bıyıkoğlu, J. Leydold, P.F. Stadler, Laplacian Eigenvectors of Graphs. Perron-Frobenius and Faber-Krahn Type Theorems (2007)
Vol. 1916: C. Villani, F. Rezakhanlou, Entropy Methods for the Boltzmann Equation. Editors: F. Golse, S. Olla (2008)
Vol. 1917: I. Veselić, Existence and Regularity Properties of the Integrated Density of States of Random Schrödinger (2008)
Vol. 1918: B. Roberts, R. Schmidt, Local Newforms for GSp(4) (2007)
Vol. 1919: R.A. Carmona, I. Ekeland, A. Kohatsu-Higa, J.-M. Lasry, P.-L. Lions, H. Pham, E. Taflin, Paris-Princeton Lectures on Mathematical Finance 2004. Editors: R.A. Carmona, E. Çinlar, I. Ekeland, E. Jouini, J.A. Scheinkman, N. Touzi (2007)
Vol. 1920: S.N. Evans, Probability and Real Trees. Ecole d'Été de Probabilités de Saint-Flour XXXV-2005 (2008)
Vol. 1921: J.P. Tian, Evolution Algebras and their Applications (2008)
Vol. 1922: A. Friedman (Ed.), Tutorials in Mathematical BioSciences IV. Evolution and Ecology (2008)
Vol. 1923: J.P.N. Bishwal, Parameter Estimation in Stochastic Differential Equations (2008)
Vol. 1924: M. Wilson, Littlewood-Paley Theory and Exponential-Square Integrability (2008)
Vol. 1925: M. du Sautoy, L. Woodward, Zeta Functions of Groups and Rings (2008)
Vol. 1926: L. Barreira, V. Claudia, Stability of Nonautonomous Differential Equations (2008)
Vol. 1927: L. Ambrosio, L. Caffarelli, M.G. Crandall, L.C. Evans, N. Fusco, Calculus of Variations and Non-Linear Partial Differential Equations. Cetraro, Italy 2005. Editors: B. Dacorogna, P. Marcellini (2008)
Vol. 1928: J. Jonsson, Simplicial Complexes of Graphs (2008)
Vol. 1929: Y. Mishura, Stochastic Calculus for Fractional Brownian Motion and Related Processes (2008)
Vol. 1930: J.M. Urbano, The Method of Intrinsic Scaling. A Systematic Approach to Regularity for Degenerate and Singular PDEs (2008)
Vol. 1931: M. Cowling, E. Frenkel, M. Kashiwara, A. Valette, D.A. Vogan, Jr., N.R. Wallach, Representation Theory and Complex Analysis. Venice, Italy 2004. Editors: E.C. Tarabusi, A. D'Agnolo, M. Picardello (2008)
Vol. 1932: A.A. Agrachev, A.S. Morse, E.D. Sontag, H.J. Sussmann, V.I. Utkin, Nonlinear and Optimal Control Theory. Cetraro, Italy 2004. Editors: P. Nistri, G. Stefani (2008)
Vol. 1933: M. Petkovic, Point Estimation of Root Finding Methods (2008)
Vol. 1934: C. Donati-Martin, M. Émery, A. Rouault, C. Stricker (Eds.), Séminaire de Probabilités XLI (2008)
Vol. 1935: A. Unterberger, Alternative Pseudodifferential Analysis (2008)
Vol. 1936: P. Magal, S. Ruan (Eds.), Structured Population Models in Biology and Epidemiology (2008)
Vol. 1937: G. Capriz, P. Giovine, P.M. Mariano (Eds.), Mathematical Models of Granular Matter (2008)
Vol. 1938: D. Auroux, F. Catanese, M. Manetti, P. Seidel, B. Siebert, I. Smith, G. Tian, Symplectic 4-Manifolds and Algebraic Surfaces. Cetraro, Italy 2003. Editors: F. Catanese, G. Tian (2008)
Vol. 1939: D. Boffi, F. Brezzi, L. Demkowicz, R.G. Durán, R.S. Falk, M. Fortin, Mixed Finite Elements, Compatibility Conditions, and Applications. Cetraro, Italy 2006. Editors: D. Boffi, L. Gastaldi (2008)
Vol. 1940: J. Banasiak, V. Capasso, M.A.J. Chaplain, M. Lachowicz, J. Miękisz, Multiscale Problems in the Life Sciences. From Microscopic to Macroscopic. Będlewo, Poland 2006. Editors: V. Capasso, M. Lachowicz (2008)
Vol. 1941: S.M.J. Haran, Arithmetical Investigations. Representation Theory, Orthogonal Polynomials, and Quantum Interpolations (2008)

Vol. 1942: S. Albeverio, F. Flandoli, Y.G. Sinai, SPDE in Hydrodynamic. Recent Progress and Prospects. Cetraro, Italy 2005. Editors: G. Da Prato, M. Röckner (2008)
Vol. 1943: L.L. Bonilla (Ed.), Inverse Problems and Imaging. Martina Franca, Italy 2002 (2008)
Vol. 1944: A. Di Bartolo, G. Falcone, P. Plaumann, K. Strambach, Algebraic Groups and Lie Groups with Few Factors (2008)
Vol. 1945: F. Brauer, P. van den Driessche, J. Wu (Eds.), Mathematical Epidemiology (2008)
Vol. 1946: G. Allaire, A. Arnold, P. Degond, T.Y. Hou, Quantum Transport. Modelling, Analysis and Asymptotics. Cetraro, Italy 2006. Editors: N.B. Abdallah, G. Frosali (2008)
Vol. 1947: D. Abramovich, M. Mariño, M. Thaddeus, R. Vakil, Enumerative Invariants in Algebraic Geometry and String Theory. Cetraro, Italy 2005. Editors: K. Behrend, M. Manetti (2008)
Vol. 1948: F. Cao, J-L. Lisani, J-M. Morel, P. Musé, F. Sur, A Theory of Shape Identification (2008)
Vol. 1949: H.G. Feichtinger, B. Helffer, M.P. Lamoureux, N. Lerner, J. Toft, Pseudo-Differential Operators. Quantization and Signals. Cetraro, Italy 2006. Editors: L. Rodino, M.W. Wong (2008)
Vol. 1950: M. Bramson, Stability of Queueing Networks, Ecole d'Eté de Probabilités de Saint-Flour XXXVI-2006 (2008)
Vol. 1951: A. Moltó, J. Orihuela, S. Troyanski, M. Valdivia, A Non Linear Transfer Technique for Renorming (2009)
Vol. 1952: R. Mikhailov, I.B.S. Passi, Lower Central and Dimension Series of Groups (2009)
Vol. 1953: K. Arwini, C.T.J. Dodson, Information Geometry (2008)
Vol. 1954: P. Biane, L. Bouten, F. Cipriani, N. Konno, N. Privault, Q. Xu, Quantum Potential Theory. Editors: U. Franz, M. Schuermann (2008)
Vol. 1955: M. Bernot, V. Caselles, J.-M. Morel, Optimal Transportation Networks (2008)
Vol. 1956: C.H. Chu, Matrix Convolution Operators on Groups (2008)
Vol. 1957: A. Guionnet, On Random Matrices: Macroscopic Asymptotics, Ecole d'Eté de Probabilités de Saint-Flour XXXVI-2006 (2009)
Vol. 1958: M.C. Olsson, Compactifying Moduli Spaces for Abelian Varieties (2008)
Vol. 1959: Y. Nakkajima, A. Shiho, Weight Filtrations on Log Crystalline Cohomologies of Families of Open Smooth Varieties (2008)
Vol. 1960: J. Lipman, M. Hashimoto, Foundations of Grothendieck Duality for Diagrams of Schemes (2009)
Vol. 1961: G. Buttazzo, A. Pratelli, S. Solimini, E. Stepanov, Optimal Urban Networks via Mass Transportation (2009)
Vol. 1962: R. Dalang, D. Khoshnevisan, C. Mueller, D. Nualart, Y. Xiao, A Minicourse on Stochastic Partial Differential Equations (2009)
Vol. 1963: W. Siegert, Local Lyapunov Exponents (2009)
Vol. 1964: W. Roth, Operator-valued Measures and Integrals for Cone-valued Functions and Integrals for Cone-valued Functions (2009)
Vol. 1965: C. Chidume, Geometric Properties of Banach Spaces and Nonlinear Iterations (2009)
Vol. 1966: D. Deng, Y. Han, Harmonic Analysis on Spaces of Homogeneous Type (2009)
Vol. 1967: B. Fresse, Modules over Operads and Functors (2009)
Vol. 1968: R. Weissauer, Endoscopy for GSP(4) and the Cohomology of Siegel Modular Threefolds (2009)
Vol. 1969: B. Roynette, M. Yor, Penalising Brownian Paths (2009)
Vol. 1970: M. Biskup, A. Bovier, F. den Hollander, D. Ioffe, F. Martinelli, K. Netočný, F. Toninelli, Methods of Contemporary Mathematical Statistical Physics. Editor: R. Kotecký (2009)
Vol. 1971: L. Saint-Raymond, Hydrodynamic Limits of the Boltzmann Equation (2009)
Vol. 1972: T. Mochizuki, Donaldson Type Invariants for Algebraic Surfaces (2009)
Vol. 1973: M.A. Berger, L.H. Kauffmann, B. Khesin, H.K. Moffatt, R.L. Ricca, De W. Sumners, Lectures on Topological Fluid Mechanics. Cetraro, Italy 2001. Editor: R.L. Ricca (2009)
Vol. 1974: F. den Hollander, Random Polymers: École d'Été de Probabilités de Saint-Flour XXXVII – 2007 (2009)
Vol. 1975: J.C. Rohde, Cyclic Coverings, Calabi-Yau Manifolds and Complex Multiplication (2009)
Vol. 1976: N. Ginoux, The Dirac Spectrum (2009)
Vol. 1977: M.J. Gursky, E. Lanconelli, A. Malchiodi, G. Tarantello, X.-J. Wang, P.C. Yang, Geometric Analysis and PDEs. Cetraro, Italy 2001. Editors: A. Ambrosetti, S.-Y.A. Chang, A. Malchiodi (2009)
Vol. 1978: M. Qian, J.-S. Xie, S. Zhu, Smooth Ergodic Theory for Endomorphisms (2009)
Vol. 1979: C. Donati-Martin, M. Émery, A. Rouault, C. Stricker (Eds.), Śeminaire de Probablitiés XLII (2009)
Vol. 1980: P. Graczyk, A. Stos (Eds.), Potential Analysis of Stable Processes and its Extensions (2009)
Vol. 1981: M. Chlouveraki, Blocks and Families for Cyclotomic Hecke Algebras (2009)
Vol. 1982: N. Privault, Stochastic Analysis in Discrete and Continuous Settings. With Normal Martingales (2009)
Vol. 1983: H. Ammari (Ed.), Mathematical Modeling in Biomedical Imaging I. Electrical and Ultrasound Tomographies, Anomaly Detection, and Brain Imaging (2009)
Vol. 1984: V. Caselles, P. Monasse, Geometric Description of Images as Topographic Maps (2010)

Recent Reprints and New Editions

Vol. 1702: J. Ma, J. Yong, Forward-Backward Stochastic Differential Equations and their Applications. 1999 – Corr. 3rd printing (2007)
Vol. 830: J.A. Green, Polynomial Representations of GL_n, with an Appendix on Schensted Correspondence and Littelmann Paths by K. Erdmann, J.A. Green and M. Schoker 1980 – 2nd corr. and augmented edition (2007)
Vol. 1693: S. Simons, From Hahn-Banach to Monotonicity (Minimax and Monotonicity 1998) – 2nd exp. edition (2008)
Vol. 470: R.E. Bowen, Equilibrium States and the Ergodic Theory of Anosov Diffeomorphisms. With a preface by D. Ruelle. Edited by J.-R. Chazottes. 1975 – 2nd rev. edition (2008)
Vol. 523: S.A. Albeverio, R.J. Høegh-Krohn, S. Mazzucchi, Mathematical Theory of Feynman Path Integral. 1976 – 2nd corr. and enlarged edition (2008)
Vol. 1764: A. Cannas da Silva, Lectures on Symplectic Geometry 2001 – Corr. 2nd printing (2008)

Editorial Policy (for the publication of monographs)

1. Lecture Notes aim to report new developments in all areas of mathematics and their applications - quickly, informally and at a high level. Mathematical texts analysing new developments in modelling and numerical simulation are welcome.

   Monograph manuscripts should be reasonably self-contained and rounded off. Thus they may, and often will, present not only results of the author but also related work by other people. They may be based on specialised lecture courses. Furthermore, the manuscripts should provide sufficient motivation, examples and applications. This clearly distinguishes Lecture Notes from journal articles or technical reports which normally are very concise. Articles intended for a journal but too long to be accepted by most journals, usually do not have this "lecture notes" character. For similar reasons it is unusual for doctoral theses to be accepted for the Lecture Notes series, though habilitation theses may be appropriate.

2. Manuscripts should be submitted either online at www.editorialmanager.com/lnm to Springer's mathematics editorial in Heidelberg, or to one of the series editors. In general, manuscripts will be sent out to 2 external referees for evaluation. If a decision cannot yet be reached on the basis of the first 2 reports, further referees may be contacted: The author will be informed of this. A final decision to publish can be made only on the basis of the complete manuscript, however a refereeing process leading to a preliminary decision can be based on a pre-final or incomplete manuscript. The strict minimum amount of material that will be considered should include a detailed outline describing the planned contents of each chapter, a bibliography and several sample chapters.

   Authors should be aware that incomplete or insufficiently close to final manuscripts almost always result in longer refereeing times and nevertheless unclear referees' recommendations, making further refereeing of a final draft necessary.

   Authors should also be aware that parallel submission of their manuscript to another publisher while under consideration for LNM will in general lead to immediate rejection.

3. Manuscripts should in general be submitted in English. Final manuscripts should contain at least 100 pages of mathematical text and should always include

   - a table of contents;
   - an informative introduction, with adequate motivation and perhaps some historical remarks: it should be accessible to a reader not intimately familiar with the topic treated;
   - a subject index: as a rule this is genuinely helpful for the reader.

   For evaluation purposes, manuscripts may be submitted in print or electronic form (print form is still preferred by most referees), in the latter case preferably as pdf- or zipped ps-files. Lecture Notes volumes are, as a rule, printed digitally from the authors' files. To ensure best results, authors are asked to use the LaTeX2e style files available from Springer's web-server at:

   ftp://ftp.springer.de/pub/tex/latex/svmonot1/ (for monographs) and
   ftp://ftp.springer.de/pub/tex/latex/svmultt1/ (for summer schools/tutorials).

Additional technical instructions, if necessary, are available on request from: lnm@springer.com.

4. Careful preparation of the manuscripts will help keep production time short besides ensuring satisfactory appearance of the finished book in print and online. After acceptance of the manuscript authors will be asked to prepare the final LaTeX source files and also the corresponding dvi-, pdf- or zipped ps-file. The LaTeX source files are essential for producing the full-text online version of the book (see http://www.springerlink.com/openurl.asp?genre=journal&issn=0075-8434 for the existing online volumes of LNM).
   The actual production of a Lecture Notes volume takes approximately 12 weeks.

5. Authors receive a total of 50 free copies of their volume, but no royalties. They are entitled to a discount of 33.3% on the price of Springer books purchased for their personal use, if ordering directly from Springer.

6. Commitment to publish is made by letter of intent rather than by signing a formal contract. Springer-Verlag secures the copyright for each volume. Authors are free to reuse material contained in their LNM volumes in later publications: a brief written (or e-mail) request for formal permission is sufficient.

Addresses:

Professor J.-M. Morel, CMLA,
École Normale Supérieure de Cachan,
61 Avenue du Président Wilson, 94235 Cachan Cedex, France
E-mail: Jean-Michel.Morel@cmla.ens-cachan.fr

Professor F. Takens, Mathematisch Instituut,
Rijksuniversiteit Groningen, Postbus 800,
9700 AV Groningen, The Netherlands
E-mail: F.Takens@rug.nl

Professor B. Teissier, Institut Mathématique de Jussieu,
UMR 7586 du CNRS, Équipe "Géométrie et Dynamique",
175 rue du Chevaleret,
75013 Paris, France
E-mail: teissier@math.jussieu.fr

For the "Mathematical Biosciences Subseries" of LNM:

Professor P.K. Maini, Center for Mathematical Biology,
Mathematical Institute, 24-29 St Giles,
Oxford OX1 3LP, UK
E-mail: maini@maths.ox.ac.uk

Springer, Mathematics Editorial, Tiergartenstr. 17,
69121 Heidelberg, Germany,
Tel.: +49 (6221) 487-259
Fax: +49 (6221) 4876-8259
E-mail: lnm@springer.com

Printing and Binding: Stürtz GmbH, Würzburg